HEINKEL
HE 177, 277, 274

HEINKEL
HE 177, 277, 274

MANFRED GRIEHL · JOACHIM DRESSEL

Airlife

England

Copyright © 1998 Manfred Griehl and Joachim Dressel

First published in the UK in 1998
by Airlife Publishing Ltd

First published in Germany in 1989 as Heinkel He 177-277-274 – *Eine luftfahrtgeschichtliche Dokumentation* by Motorbuch Verlag, Stuttgart.

British Library Cataloguing-in-Publication Data

A catalogue record for this book
is available from the British Library

ISBN 1 85310 364 0

All rights reserved. No part of this book may be reproduced or transmitted in any form or by any means, electronic or mechanical including photocopying, recording or by any information storage and retrieval system, without permission from the Publisher in writing.

Typeset by Phoenix Typsetting, Ilkley, West Yorkshire.
Printed in Hong Kong

Airlife Publishing Ltd
101 Longden Road, Shrewsbury, SY3 9EB, England

Foreword

Heinkel's Project P 1041, forerunner of what became the He 177, was conceived by Siegfried Günter who developed his idea under the technical direction of Dipl-Ing Heinrich Hertel. As a long-range 'Big *Stuka*', the He 177 was to be as it were one of the pillars of the planned *Luftwaffe*'s new armament. But what actually became of Prof Dr-Ing Ernst Heinkel's vision of this new aircraft being used primarily as an advanced long-range bomber?

Was the He 177, as noted by Karl-Albin Kruse, 'at the end of its development the most perfect bomber of its day'? Was it really, according to Hans Redemann, 'revolutionary in its technical conception, and years ahead of other similar developments'? And is his claim correct that only the bureaucratic narrow-mindedness and lack of determination on the part of the leaders at the top hindered the He 177, with its novel technology and configuration, from becoming 'a great success'?

After Dipl-Ing Hertel had fallen out with Prof Dr-Ing Heinkel and left the company, and his successor, Dipl-Ing Robert Lusser, one of the principal protagonists of the 'diving He 177', had been given his notice, the Heinkel works policy at least had been consolidated. The subsequent appointment of Dipl-Ing Hertel by the RLM as '*Commissar* for the He 177' was to highlight his industrious endeavours to eliminate the aircraft's shortcomings; but Heinkel's former Technical Director and Chief of Development could not – or would not – eliminate the handicaps imposed by the RLM on the design and development of the He 177.

The whole story of the He 177 accordingly represents an almost mirror image of a misplaced and misguided air armament policy, in many ways affected by personal aspects, the aggravating influence of which could not fail to have a detrimental effect on the He 177's later service use.

We would like to express our thanks to the following for their generous support and help in the form of words and illustrations: U. Balke; P. Zollner of the Bayerischen Motoren Werke AG; H. Birkholz; R. Chapman; E. Creek and the Air Archives of the Deutsches Museum München, especially Dr Heinrichs and Dr Limmer; also H.-P. Dabrowski; K. Francella; *Herr* Meier of Frankfurt Airport; P. Heck; P.K. Herrendorf; D. Herwig of the Deutsches Studienbüro für Luftfahrt; Dr A. Hiller; K. Junges; Dipl-Ing K. Kössler; H. Kruse; B. Lange; H.J. Meier; H. Roosenboom, and E. Götsch and H. Pause of the Henschel-Werke GmbH.

We also owe our thanks to R.P. Lutz Jr; F. Marshall; J. Menke; F. Müller-Romminger; H.J. Nowarra; R. Mikesh and Mrs S.E. Ewing of the National Air & Space Museum in Washington, DC; P. Petrick; W. Pervesler; Dr A. Price; W. Radinger; Ch. Regel; H. Riediger; Siemens AG (Munich); F. Selinger; H. Schliephake; R. Schirer; R. Smith; K. Soppa; H.H. Stapfer; H. Thiele; F. Trenkle and F.A. Vajda. Likewise to the Bundesarchiv/Militärarchiv and *Herren* Albinus and Nilges for their friendly support, as well as the Wehrbereichsbibliothek IV at Mainz.

Our special thanks are due to *Herr* K.-H. Heinkel for allowing us to look through the records of the former Heinkel AG, which provided extremely valuable help to the authors.

Finally, we would like to express our thanks to *Frau* M. Junges for checking the manuscript, and to *Frau* Dr Scholten-Pietsch of Motorbuch Verlag for her readiness to publish this manuscript.

Manfred Griehl & Joachim Dressel
Mainz, Hochheim

Luftwaffe Formations:
An Explanatory Note by the Translator

Although a wartime *Luftwaffe* bomber *Staffel* seems similar to an RAF or USAAF squadron, it generally had fewer aircraft on strength and differed in its subordination, while the larger *Gruppe* and *Kampfgeschwader* formations varied in size and establishment and had no exact equivalents in the wartime RAF and USAAF. For these reasons, this English-language edition has maintained the original *Luftwaffe* unit designations.

Depending on its role, a *Staffel* could have an establishment of between 8-14 aircraft (including reserves), although from 1942 onwards the average strength of a bomber *Staffel* was seldom more than 7-10 aircraft, including those undergoing repairs. The basic operational unit was the *Gruppe*, generally comprising three *Staffeln*. A *Kampfgeschwader* had a nominal strength of three (I-III) *Gruppen*; the fourth (IV) *Gruppe* was usually a conversion/operational training unit. A few *Kampfgeschwader* were temporarily expanded to include a fifth (V) *Gruppe*, while others never increased beyond I *Gruppe*, e.g. I/FKG 50.

There were also many separate *Kampfgruppen*, some of only temporary existence, while the *Gruppen* (and sometimes even individual *Staffeln*) of a *Kampfgeschwader* could be deployed in different combat zones, far away from their *Stab* (Headquarters Flight). All *Staffeln* within a *Kampfgeschwader* were numbered in consecutive order, this number also indicating their *Gruppe*. In the majority of cases, each *Gruppe* comprised three *Staffeln*.

As a rule, the formation designations were used in abbreviated form where the *Staffeln* within a *Kampfgeschwader* were identified by Arabic numerals and the *Gruppen* by Roman numerals. Thus, 6./KG 40 was 6. *Staffel* (part of II *Gruppe*) of *Kampfgeschwader* 40; III/KG 1 was III *Gruppe* (comprising 7., 8. and 9. *Staffeln*) of *Kampfgeschwader* 1.

For specific tasks a number of such formations were organised into a *Fliegerdivision* (identified by Arabic numerals), while a *Fliegerkorps* (Roman numerals) represented an operational parallel of a *Luftgau* (Air Zone), an administrative organisation within a set area of Germany, and usually made up of a mixture of fighter, bomber and other flying units. Several *Fliegerkorps* and various other formations would make up a *Luftflotte* (Air Fleet), the strength of which varied according to operational requirements.

Contents

1. The He 177 is born   1
The Way to War
Project P 1041
Slow Progress
Cleared to Fly
Type Trials and the He 177A-0
Defensive Armament
Trials with Drop-Loads

2. The Start of operational service   50
The Way to the He 177A-1
Operational Trials and *E-Staffel* 177
He 177 Production
The He 177 as a Training Aircraft
Initial Operations of I/FKG 50
Operations of KG 1 'Hindenburg'

3. The He 177 in maritime warfare   92
From He 177A-3 to 'Spring Bomber'
The He 177 as a Torpedo-Bomber
The He 177 *Zerstörer*
He 177 Trials with Guided Weapons
The He 177 in Service with KG 40
The He 177 in Service with KG 100 'Wiking'
The He 177 in Service with *Wekusta/ObdL*

4. The Long-range bomber without a chance   157
The RLM Demands Four Separate Engines
Tests and Trials to the Last Day
Final Employment: The *Mistel*

5. The He 274 and He 277 strategic bombers   176
The He 177 High-Altitude Bomber: A Wasted Chance
Development Task: 8-274
The '*Amerika Bomber*'
He 274 versus He 277
War Booty

Appendices   220
Bibliography   242
Photo Credits   242
Glossary   243
Index   246

Chapter 1
The He 177 is Born

The Way to War

1935 was an important year in the development of the *Luftwaffe*, Germany's fledgling air force, the existence of which was revealed to the world by Adolf Hitler on 9 March that year. An independent branch of the armed forces subordinated to the Chief of the *Oberkommando der Wehrmacht* (OKW: Armed Forces High Command), General Werner von Blomberg, and built up under conditions of great secrecy and with much ingenuity (the strict Air Clauses of the Treaty of Versailles, signed in 1919 in the aftermath of the First World War, expressly forbade Germany from establishing an air force and producing military aircraft), the *Luftwaffe* already possessed some 1,800 aircraft and 20,000 men three months prior to its official unveiling.

The responsibility for overseeing the production of new military aircraft and the expansion of the new air force was that of General Hermann Göring, *Oberbefehlshaber der Luftwaffe* (ObdL: Commander-in-Chief Air Force) and a long-time close political ally of Hitler. His appointment seemed fitting, given his record as an ex-fighter pilot with 20 'kills' and final commander of the famed '*Richthofen Jagdgeschwader*' during the First World War. But while Göring concentrated on the politics, it was his deputy, State Secretary of Aviation Erhard Milch, a First World War fighter pilot who had worked in Germany's booming commercial aviation sector during the post-war years, who effectively ran the *Reichsluftfahrtministerium* (RLM: State Ministry of Aviation) in the 1930s and tried to convert Hitler's and Göring's plans for expansion of Germany's air power into realistic production programmes for Germany's aircraft industry.

Milch's efforts soon began to bear fruit, with new aircraft being built and delivered in quite substantial numbers. Of 4,021 military aircraft ordered for delivery by 30 September 1935 (including 1,863 combat types and 1,760 trainers), 2,105 had been delivered by January of that year; an average of 162 per month.

Hitler's statements and actions, however, soon saw Milch's production programme superseded by new and higher production targets, with greater emphasis on combat types. The average monthly delivery figures increased to 190 aircraft in the first half of 1935, then to 300 by the end of the year. But the demand for ever greater monthly deliveries failed to take into account the fact that Milch's carefully calculated 1934-35 production programme was based on the limited production capacity of Germany's relatively small aircraft industry, and a long-term plan formulated by General Göring; a plan that forecast the *Luftwaffe* would not reach the peak of its operational effectiveness until 1943. The significance of this schedule would become apparent in 1936, when Hitler announced a second Four-Year Plan. The first Four-Year Plan, announced in April 1933, aimed to rebuild the stagnant German economy and thus reduce unemployment in what was dubbed the 'Battle for Work'. The aim of the second plan was far more militaristic in nature: prepare Germany's armed forces and economy for war within four years.

For the *Luftwaffe*, 1936 was a milestone in terms of the development of new aircraft and the realistic evaluation of their combat capabilities. Pre-production examples of several new combat types such as the Bf 109 fighter and Hs 123 single-seat and Ju 87 two-seat dive-bombers were undergoing service trials at *Erprobungsstelle* (E-Stelle: Proving/Test Centre) Rechlin. Dornier and Heinkel were also represented in the shape of the Do 17 and He 111, destined to enter service with the *Luftwaffe* as part of a new generation of twin-engined medium bombers.

That new and better bombers were sorely needed by the *Luftwaffe* was beyond question. When the air force was revealed to the world in early 1935, its bombing component amounted to just five *Staffeln* (Squadrons) in I/KG 154 'Boelcke' and I/KG 252 'Hindenburg', their primary equipment being the Ju 52/3m ge: a makeshift bomber conversion of the Ju 52/3m 17-passenger commercial transport designed by *Diplomingenieur* (Dipl-Ing: academically-qualified engineer) Ernst Zindel that was itself a triple-engined development of the single-engined Ju 52/1m.

Though popular with its crews, the fact that this sturdy trimotor from Dessau was in the majority in the admittedly small bombing force was an indication of the problems being experienced with the *Luftwaffe*'s intended primary bomber, the Do 11C. Developed by Dornier's facility at Altenrhein in Switzerland as the Do F (the military designation was adopted in 1933), and first flown on 7 May 1932, this ungainly twin-engined design featured a bomb-bay in an angular fuselage beneath the slab-like shoulder-mounted wing. Maximum speed was a respectable 250 km/h (155 mph), range 1,200 km (746 miles).

Ordered into mass production in late 1932, the re-engined Do 11C entered military service the following year with the *Behelfsbombergeschwader* (*BehBG*: Auxiliary Bomber Group), established in October 1933 to provide the nucleus of an operational bombing force. Because the build-up of the *Luftwaffe* was highly secret, the *BehBG* adopted the cover title of *Verkehrsinspektion der DLH* (Traffic Inspectorate of Deutsche Lufthansa).

The civilian guise was enhanced in November 1933, when the *Deutsche Reichsbahn* (German State Railway) replaced two of its night-train services with corresponding air freight services. Ostensibly operated by Deutsche Lufthansa (DLH: German State Airline), the services were actually flown by Do 11Cs crewed by personnel seconded from *BehBG* 1, the flights being used to provide valuable experience in the skills of navigation and night-flying. The Do 11Cs bore no signs of military intent, their gun positions being covered and a solid nose fitted; but each aircraft's spares supply included three Rheinmettal 7.92 mm MG 15 machine-guns and corresponding nose, dorsal and ventral mountings, an interchangeable nose section featuring glazing to aid the bombardier, and racks to hold up to 1,000 kg (2,205 lb) of bombs.

Unfortunately the Do 11C was plagued by poor handling characteristics and structural defects, and production lagged far behind the 372 examples ordered for delivery during 1934. The result, in March of that year: the renamed *Behelfskampfgeschwader* (*BehKG*: Auxiliary Bomber Group) 1, joined during that year by *BehKG* 172, had just three Do 11s on strength, operating alongside 24 Ju 52/3m ges, the latter type having been acquired for service as a bomber pending the arrival in quantity of the Do 11C.

In reality, the Do 11C's problems were serious enough to warrant it being replaced by a new variant, the Do 11D, 76 of which were in service by the end of 1934. But new problems arose, and orders outstanding for 222 examples were transferred to a new model, the Do 13C, the prototype of which had flown as far back as 13 February 1933. A fixed, spatted main undercarriage replaced the Do 11C/D's temperamental retractable units, and 'double-wing' flaps were fitted to improve handling. However, the loss of several early aircraft due to structural failure hindered the Do 13C's introduction into service.

By the end of 1935 the now-official *Luftwaffe*'s bomber force was established in several *Kampfgruppen* (Bomber Wings), two-thirds of their *Staffeln* being equipped with Ju 52/3m ges or more powerful g3es, capable of carrying a 1,500 kg (3,307 lb) bomb-load in three bomb-bays. Defensive armament comprised two MG 15 machine-guns; one in an open dorsal position, the other in a semi-retractable ventral 'dustbin' attached to a fixed underfuselage fairing used by the bomb-aimer.

The remaining *Staffeln* flew dwindling numbers of Do 11Ds alongside the newer Do 23F and more powerful Do 23G, these being Do 13Cs fitted with a restressed, fabric-covered fuselage, modified wing and more powerful engines. A total of 210 Do 23s were delivered, but their service life was short. One year later, Ju 52/3m bombers, 760 of which were delivered, still equipped two-thirds of the *Luftwaffe*'s bomber force; the equivalent of eight out of the 12 *Kampfgruppen* (24 *Staffeln* with 12 aircraft each) by then established.

As the Do 23F/Gs began to undertake second-line duties, their place was taken by two new designs, one each from Junkers and Heinkel, built in response to a 1934 specification issued jointly by the C-Amt (Technical Department) of the *Luftfahrtkommissariat* and Lufthansa (the airline's name was shortened from Deutsche Lufthansa on 1 January 1934). The specification called for a high-speed, twin-engined aircraft to be used as a bomber by the clandestine *Luftwaffe* and as a 10-passenger commercial transport by Lufthansa on its *Blitz-Strecken* (Lightning Routes), the latter initially linking Berlin, Hamburg, Cologne and Frankfurt am Main.

Junkers' answer to the specification was the Ju 86, first flown at Dessau on 4 November 1934 and notable for its use of flush-riveted stressed skinning and an oval-section fuselage in place of the company's familiar corrugated sheet-metal skinning and angular fuselage.

Following evaluation of pre-production Ju 86A-0s by *E-Stelle* Rechlin, production series Ju-86A-1s entered service with KG 152 'Hindenburg' in mid-1936. The bomber could carry up to eight 100 kg (220 lb) SC 100 bombs, their release activated by the bombardier located in the aircraft's heavily-glazed nose. Three MG 15 machine-guns were carried; one each in the nose, a dorsal turret and a rotatable, semi-retractable ventral unit.

The A-1 was quickly superseded by the D-1, with increased fuel capacity and improved handling; but both models were hampered by the poor performance of their Jumo 205C-4 six-cylinder diesel engines. A solution was found in part by switching to BMW 132F nine-cylinder radials, but the earlier problems led to the cutting back of plans to equip 12 *Kampgeschwaders* (KG: Bomber Groups) with Ju 86s as part of the 1937 expansion plan drawn up for the fast-growing *Luftwaffe*.

Heinkel's response to the C-Amt/Lufthansa specification was the He 111. First flown on 24 February 1935, it was more elegant in appearance than the Ju 86, but the two passenger compartments were too cramped and its overall performance rendered it uneconomical for use as a commercial transport. The BMW VI6,0Z-powered pre-production He 111A-0 bomber was also rejected for service, after trials at *E-Stelle* Rechlin showed it to be underpowered when fully-loaded. Heinkel, aware of the problem, responded by adopting the more powerful Daimler-Benz DB 600C 12-cylinder liquid-cooled engine. The improvement in performance over that of the BMW-powered A-0 was dramatic: maximum speed rose from 309 km/h (192 mph) to 360 km/h (224 mph); maximum cruising speed when fully-loaded from 270 km/h (168 mph) to 339 km/h (211 mph).

Like the Ju 86, the He 111 had a four-man crew, with the bombardier housed in an extensively glazed nose that also housed one 7.7 mm machine-gun. Two MG 15s were also carried; one in a dorsal turret, the other in the ubiquitous semi-retractable ventral 'dustbin' used by the radio-operator. The bomb-load, typically eight 100 kg (220 lb) SC 100s, was stowed vertically, nose-up.

The He 111B-1 began to enter service with KG 154 'Boelcke' in the winter of 1936-37, several months after the start of the Spanish Civil War in which Right-wing army officers, fearful that the weak liberal government of the Centre was about to be pushed aside in a full-scale revolution orchestrated by Communists in Spain and the Soviet Union, conspired to seize power from the Republicans. Outside intervention was swift, Germany and Italy siding with the Nationalists while the Soviet Union backed the Republican cause.

All three nations contributed military personnel and hardware, including large numbers of combat aircraft. Germany's tangible support began within a week of the outbreak of war on 18 July 1936, with the despatch of 20 Ju 52/3m g3e transport-bombers to Seville, their primary role being to help airlift some 14,000 Army of Africa troops, commanded by General Francisco Franco y Bahamonde, across the Strait of Gibraltar from Spanish Morocco to southern Spain as a prelude to a northwards advance along the Tagus Valley towards Toledo and then the nation's capital, Madrid.

In Germany, General Göring and the *Luftwaffenführrungsstab* (Air Force Operations Staff) saw the Spanish Civil War as a useful proving ground on which to evaluate new combat aircraft and tactics under operational conditions, as well as providing valuable combat experience for *Luftwaffe* personnel. Consequently, Göring despatched six He 51B-1 biplane fighters for the Nationalists, along with a strong contingent of 85 'volunteers' of the so-called *Reisegesellschaftsunion* (Tourist Group) who would staff a *Luftwaffe* training unit in Spain. Heavily disguised as *Luftübung Rügen* (Air Exercise *Reprimand*), the aircraft and personnel left Hamburg for Cádiz on 31 July as the first stage in the secret build-up of the *Luftwaffe*'s Legion Condor, established to provide air support for Nationalist ground forces.

Despite furious foreign reaction, this highly-capable and semi-autonomous air component, commanded by *Generalmajor* Hugo Sperrle (later replaced by *Generalmajor* Helmuth Volkmann) with *Oberst* Wolfram von Richthofen (a cousin of Manfred von Richthofen, the famous First World War fighter pilot) as Chief-of-Staff, was resolutely enlarged.

By 6 November 1936 the Legion Condor was ready for action, and made its combat debut just 11 days later. For the duration of the war, *Luftwaffe* personnel were assigned to the component *Staffeln* on a roster basis, returning home and passing on first-hand their combat experience to the *Luftwaffe*'s flying training schools and their students. Such information was vital for Göring, who already seemed to envisage being confronted by a much larger conflict:

> 'We are already at war, although no shots are being fired as yet. The situation is very serious: Russia wants war, and England is rearming strongly.'*

The Condor Legion's bombers were assigned to *Kampfgruppe* 88 (K/88), initially comprised of three 12-aircraft *Staffeln* equipped with Ju 52/3m g3es and g4es. In February 1937 3.K/88 and the new 4.K/88 received the first of 30 He 111B-1s sent to Spain. Less than a month later, on 9 March, He 111B-1s undertook the type's first operational sortie: an attack on Republican airfields at Alcalá and Barajas. Subsequent experience revealed the Heinkel bomber to be fast enough to elude the majority of Republican fighter types, and able to carry a short-range bomb-load of 1,500 kg (3,307 lb). Attrition was low on what were usually unescorted daylight raids, so 1. and 2.K/88 also received He 111B-1s and the more powerful B-2. All four *Staffeln* went on to receive a total of 45 He 111E-1s in 1938, this Jumo 211A-1-powered model having the ability to carry a 2,000 kg (4,409 lb) bomb-load.

The operational effectiveness of the He 111 overshadowed that of its Junkers counterpart, the Ju 86. Five D-1s were assigned to K/88 in late 1937; but the proving ground that was the Spanish Civil War found them wanting, their engines proving particularly vulnerable to the stresses and strains of combat flying.

In contrast, another new bomber, the Dornier Do 17, was judged a success in the initial stages of its deployment to Spain. Designed as a commercial transport to meet Lufthansa's need for a high-speed mailplane able to carry six passengers, the 'Flying Pencil' (so-called because of its long, slim fuselage) was rejected by the airline (the passenger cabins were deemed to be too cramped), then reborn as a bomber for the *Luftwaffe*.

The first production series model was the Do 17E-1 bomber, capable of carrying a 750 kg (1,653 lb) bomb-load and which entered *Luftwaffe* service in early 1937, followed closely by the reconnaissance-configured F-1. Both variants were in service with the Legion Condor by mid-1937, 20 E-1s being sent to supplement the He 111B-1/2s of 1. and 2.K/88, while 15 F-1s from the home-based *Fernaufklärungsgruppe* (AufklGr(F): Long-Range Reconnaissance Wing) 122 formed 1.A/88 to replace the *Staffel* of He 70F-2s reconnaissance-bombers that formed the Legion's A/88.

The combat conditions experienced by Do 17E/F-1s in the early stages of the deployment revealed the Dornier design to be virtually immune to interception by Republican aircraft. However, as the war progressed and the Republicans acquired new and better fighters from the Soviet Union, the Do 17s found themselves increasingly vulnerable. Defensive fire on both models comprised just two MG 15s: one aft-firing from the rear of the flight deck, the other downward-firing through a ventral hatch. Ten Do 17P-1s sent to supplement the F-1s sported a third MG 15, this one a forward-firing weapon operated from the cockpit by the pilot. But combat experience had highlighted the aircraft's belly as being particularly vulnerable to attack, the single ventral MG 15 having too limited a field of fire for effective defence.

With the fall of Valencia and Madrid to the Nationalists in late March 1939, followed almost immediately by international recognition of the government of General Franco, the Spanish Civil War finally came to an end. The Legion Condor was formally disbanded in April; but interpretation of its units' combat experience, long the subject of intense scrutiny within the OKL, would continue to influence the young *Luftwaffe*'s fighting capabilities, in particular the nature of its bombers and their operational capabilities, for years to come.

---

* File note re. discussions on 2 December 1936; General Karl Bodenschatz.

Crucially, General Göring ignored the Legion's success when used in a strategic role, namely the bombing of Republican ports, and concentrated instead on its undoubted and spectacular success as a ground-support force operating in the tactical role, first with bomb-carrying He 51B-1s and then with Hs 123A-1s and Ju 87A/B-1s operating in the close-support and dive-bombing roles respectively. Thus was born the *Blitzkrieg* (Lightning War): short, sharp tactical onslaughts prefaced by an overwhelming initial blow using a deadly combination of surprise and highly-mobile firepower to knock out the enemy's forces before they had time to mount any credible defence.

The adoption of this form of warfare had a significant impact on Germany's rearmament programme and the nature of the weaponry to be produced, with the emphasis now very much on quantity to ensure short-term overwhelming superiority and thus enable a quick victory. For the *Luftwaffe*, whose primary role was seen as tactical close-support of the army rather than strategic offensives, that meant large numbers of dive-bombers and medium bombers instead of long-range heavy bombers. In simple terms, three twin-engined medium bombers could be built for every two four-engined heavy bombers.

Göring's advocacy of a *Luftwaffe* optimised for tactical operations in support of ground forces sounded the death-knell for two designs being developed by Dornier and Junkers to meet a perceived requirement for a four-engined 'Ural Bomber' capable of carrying an effective bomb-load to the Ural Mountains (the eastern border of Hitler's much-vaunted *Lebensraum* (Living Space): land that would be conquered to enable the establishment of a Germanic empire).

The concept of a *Langstrecken-Grossbomber* (Long-Range Heavy Bomber) originated within the *Luftwaffenführrungsstab* in 1934, and received the support of *Generalleutnant* Walther Wever, the *Luftwaffe*'s first Chief-of-Staff and an enthusiastic proponent of strategic bombing. Wever's backing led the RLM's *Technische Amt* (Technical Office) to issue a specification for such an aircraft to Dornier and Junkers in the summer of 1935.

Dornier's response was the inelegant Do 19, characterised by a slab-sided fuselage, thick low-mid wing, and braced vertical tail surfaces mounted atop the tailplane. A full-scale mock-up was followed by the Do 19 V1 (D-AGAI/Wk-Nr 701), first flown on 28 October 1936. Powered by four Bramo (Siemens) 322H-2 nine-cylinder air-cooled radial engines, the V1 was unarmed. Two more prototypes were scheduled: the V2

The unarmed Do 19 V1 (D-AGAI/Wk-Nr 701), Dornier's four-engined response to a mid-1930s request for a long-range heavy bomber.

Junkers' response to the same request was the Ju 89, illustrated by the Ju 89 V1 (D-AFIT/Wk-Nr 4911), seen at Dessau during factory tests.

(Wk-Nr 702) with the more powerful BMW 132F nine-cylinder radials; and the V3 (Wk-Nr 703), the first example to feature the planned defensive armament of one 20 mm MG FF cannon in each of two two-man hydraulically-powered turrets (one dorsal, one ventral) and two MG 15 machine-guns (one in a nose turret operated by the bombardier, the other in an open tail position).

Static tests revealed the dorsal and ventral turrets to be too heavy, to the extent that the structural strengthening of the fuselage deemed necessary would have increased the Do 19's all-up weight to such an extent that the aircraft would have been left seriously underpowered. More powerful engines and lighter turrets were proposed for the production-standard Do 19A, the belief being that these changes would endow the aircraft with a maximum speed of 370 km/h (230 mph) and a maximum range of 1,995 km (1,240 miles). The 1,600 kg (3,527 lb) bomb-load would consist of 16 100 kg (220 lb) SC 100s or 32 50 kg (110 lb) SC 50s.

The Junkers Ju 89 was also designed to carry a 1,600 kg (3,527 lb) bomb-load, and a crew of nine that included five gunners. All-metal, the aircraft featured a flush-riveted stressed duralumin skin, twin outboard tailfins, Junkers' 'double-wing' flaps, and hydraulically-operated main undercarriage units that retracted into the rear of each inboard Jumo 211A 12-cylinder liquid-cooled inline engine nacelle.

Three prototypes were to be built, the unarmed V1 (D-AFIT/Wk-Nr 4911) making its maiden flight in December 1936, followed by the equally unarmed but DB 600A-powered V2 (D-ALAT/Wk-Nr 4912) in early 1937. As with the Do 19 V3, the Ju 89 V3 (Wk-Nr 4913) was to evaluate the type's defensive armament: two 20 mm MG FFs in each of two two-man hydraulically-operated turrets (again one dorsal, one ventral), and two MG 15s (one firing through the windscreen, the other from a tail position). Maximum speed of the unarmed V1 was established at 389 km/h (242 mph); maximum range at 1,995 km (1,240 miles).

Even before the first prototypes of each aircraft had taken to the air, the entire 'Ural Bomber' project suffered a major blow with the death in an air crash on 3 June 1936 of *Generalleutnant* Wever, the *Luftwaffe*'s most ardent advocate of strategic bombing. His replacement, in a post now known as Chief of the *Luftwaffengeneralstab* (Air Force General Staff), was *Generalleutnant* Albert Kesselring, who believed that the *Luftwaffe*'s primary role in any war in Western Europe would be tactical rather than strategic, and therefore production of fighters and medium bombers should be given priority over long-range heavy bombers.

The Do 19 V1 and Ju 89 V1 and V2 were all engaged in flight-testing when the 'Ural Bomber' project was officially cancelled by the RLM on 29 April 1937. Protestations by the *Technische Amt* to General Göring went unheeded; for, as Göring himself had pointed out to his confidantes earlier that year:

'The Führer does not ask me how big my bombers are, but how many I have.'*

The Do 19 V2 and V3 were scrapped before completion, but the V1 was modified as a transport and went on to see service with the *Luftwaffe* during the Polish campaign of 1939. As for the Ju 89s, the V1 and V2 continued flight-testing after the cancellation of the 'Ural Bomber' project, but in support of development of a commercial derivative, the Ju 90, before being converted as transports and assigned to the *Luftwaffe*'s *Kampfgruppe zur besonderen Verwendung* (KGrzbV: Battle Wing for Special Duties/Transport Wing) 105 during the invasion of Norway in 1940.

The Ju 89 V3 was cannibalised before completion, its wings, undercarriage, tail unit and engines being used in the construction of the Ju 90 V1 (D-AALU/Wk-Nr 4913) transport, first flown on 28 August 1937 and followed in 1938 by the V2 (D-AIVI/Wk-Nr 4914) and V3 (D-AURE/ Wk-Nr 4915).

The Ju 90 V4 (D-ADLH/Wk-Nr 4916) was built as the prototype of the B-1 38-40 passenger commercial transport, 10 of which (Wk-Nrn 90 0001-10) were set for service with Lufthansa (eight) and South African Airways (two, designated Z-2). A few did enter service with Lufthansa, but these plus the two Z-2s were pressed into *Luftwaffe* service in 1939, going on to serve with *Lufttransportstaffel* (LTS: Air Transport Squadron) 290. Two were eventually returned to Lufthansa, but others were used in the development of the Ju 290A-series large-capacity transport/maritime reconnaissance-bomber (*see* Chapter 5).

The fact that the Secretary of State for Air, Erhard Milch, heard of General Göring's decision to cancel further development of the 'Ural Bomber' (following a discussion between Göring, *Generalleutnant* Kesselring and *Oberst* Hans Jeschonnek of the *Luftwaffengeneralstab*) only by chance, spoke volumes for the relationship between Göring, vain and tempestuous and Milch, cool and efficient. As the working climate between the two grew steadily worse, the 'Iron Man' Göring began to have most development discussions directly with departmental chiefs, frequently excluding Milch altogether.

At the same time Göring's *protégé*, the 'less dangerous' *Oberst* Ernst Udet, a highly-experienced First World War fighter pilot and one of the founders in 1922 of the *Udet-Flugzeugbau* GmbH, was pushed further into the limelight (at the expense of Milch), being reassigned in June 1936 from Inspector of Fighter and Dive-Bomber Pilots to concern himself with the technical development of *Luftwaffe* equipment as Chief of the Development Section of the *Technische Amt*.

In November 1936, the year in which he joined the *Luftwaffe* having previously juggled his responsibilities at the RLM with his role as Chairman of Lufthansa, Erhard Milch asked to be relieved of his duties as State Secretary for Air. His request was turned down. Udet, meanwhile, was becoming more and more a means to an end for Göring; being used to strongly enhance the General's reputation (and self-esteem) concerning the build-up of the *Luftwaffe*, and to belittle the obvious merits of Milch. In this way, everything was done to sow mistrust and envy between Göring and Milch, as well as between the operative command and the technical leadership of the future.

Ironically, Göring and Milch were in general agreement about the use of the *Luftwaffe* as a tactical force, but for different reasons. Milch opposed development of the 'Ural Bomber', arguing that the production of a relatively small number of long-range heavy bombers did not justify what he saw as too great a challenge for the German aircraft industry in the mid-1930s. He believed the emphasis should be on the production of twin-engined bombers and dive-bombers – but over a realistic period of time rather than the quick build-up of the *Luftwaffe* being demanded by Göring and Hitler; a goal that in Milch's opinion was bound to fail when faced by the reality of the German aircraft industry's still-limited production capacity.

To reduce Milch's existing influence still further, in May 1937 Göring divided the RLM into two branches – one at ministerial level, the other the domain of *Generalleutnant* Kesselring – both of

---

* File note dated 22 Jan 1937.

which he intended to run himself. But that still wasn't enough. Ultimately, he also branched off the Personnel Department under General Robert Ritter von Greim, a wartime compatriot of Ernst Udet and the first squadron leader of the new *Luftwaffe*; while the *Technische Amt* became the responsibility of Udet himself. In this way, all four department heads – Kesselring, Milch, Udet and Ritter von Greim – were accorded equal status; and so Milch's scope of influence was reduced as never before. The result was an end to practical co-operation between departments, making all efforts to build a four-engined long-range heavy bomber all but impossible.

The long-term effect of all the political infighting and petty jealousies was best summed up in late 1942 by the President of the State Association of the German Aviation Industry, Admiral Lahs, who is reported to have stated in a note to *Generalfeldmarschall* Milch:

> 'In 1936, Junkers and Dornier had both built prototypes of heavy bombers. With systematic progressive development, they would have been today, six years later, superior to all American and British long-range bombers.'*

Project P 1041

Although the death of *Generalleutnant* Wever undoubtedly robbed heavy bomber development in Germany of one of its most important advocates, which in turn effectively marked the beginning of the triumphant progress within the *Luftwaffe* of the medium bomber capable of diving attacks, the loss of his support was not responsible, either directly or indirectly, for the cancellation of long-range strategic bomber development in Germany. In fact, some months before the Do 19 V1 conducted its maiden flight on 28 October 1936, the *Luftwaffenführungsstab*, having realised that the anticipated performance of both 'Ural Bomber' designs had been overtaken by a requirement for a more advanced heavy bomber with more exacting performance parameters, formulated a new set of requirements in what was known as the 'Bomber A' specification.

---

* GL file note, undated

On 3 June 1936 the leading representatives of Blohm und Voss, Heinkel, Henschel, Junkers and Messerschmitt were instructed by the RLM to start planning a heavy bomber capable of diving attacks. All 'Bomber A' project studies were to be presented by August 1936, just two months later, the object being to create a three-seat long-range bomber equipped with effective defensive armament.

According to the specification issued by the *Technische Amt*, the resulting operational aircraft had to be capable of a maximum speed of 500 km/h (311 mph) and have a maximum range of 5,000 km (3,107 miles). The following engines could be used:

- Argus 421   24-cylinder air-cooled H-type
- BMW 139   14-cylinder air-cooled radial
- DB 601   12-cylinder liquid-cooled inline
- Jumo 206   6-cylinder liquid-cooled inline
- Jumo 211   12-cylinder liquid-cooled inline
- SAM 329   14-cylinder air-cooled radial

The number of engines used was left to the manufacturers' discretion. General-Ing Lucht, Chief of the Engineering Division of the *Generalluftzeugmeister-Amtes* (Chief of Aircraft Procurement and Supply Offices), also gave a real opportunity to such solutions as remote-drive, bell-crank or angle-drive engines, as well as engines mounted inside the fuselage that could be accessible in flight.

The new bomber had to make do with a take-off run of 1,000 m (3,281 ft) or – in overloaded conditions – get airborne with the assistance of a catapult. General-Ing Lucht also placed special emphasis on operational tactical aspects: in addition to the usual 10, 50 and 250 kg (22, 110 and 551 lb) bombs, each project study had to guarantee that its new bomber could at any time also take bombs of larger calibre.

The three-man crew was to act as four. The aircraft was to be flown by two pilots, one of whom would act as commander and simultaneously operate one of the defensive weapons. The third crewman was the radio operator who was also a gunner. For the time being, no great value was attached to the use of a pressurised cockpit: at an operational altitude of over 6,000 m (19,685 ft) the crew would in any case be wearing 'diver suits' with oxygen supply.

With several remote-controlled 13 mm MG 131

machine-gun installations, the defensive armament was intended to provide sufficient protection against enemy fighters. The new bomber was also to feature a heavy automatic weapon for air-to-ground use. In early 1937, during discussions with Dipl-Ing Heinrich Hertel, the Ernst Heinkel Flugzeugwerke's Technical Director and Chief of Development, concerning the company's Project P 1041 response to the 'Bomber A' specification, General-Ing Lucht welcomed Hertel's suggestion not to arm the new bomber with fixed machine-guns or cannon in the cockpit; but instead to fit the weapons in the wing roots, to ensure the optimum use of space in the proposed relatively narrow cabin and thus achieve a higher maximum speed.

Dipl-Ing Hertel even proposed a long-range bomber where all three crewmen would be seated in tandem. This particular aircraft was to have a range of 2,500 km (1,553 miles) with a maximum bomb-load of 2,200 kg (4,850 lb). Thanks to the simplest possible layout and the size of the fuselage, it was intended to create a long-range bomber that would be above all cheap to produce and weigh only a little more than a medium bomber. In addition, the kind of construction envisioned offered all the advantages of a long-range bomber.

The new bomber's entire armament was later to be remote-controlled, but until then the experimental aircraft (according to a proposal submitted to Ernst Udet early in December 1937) were to be fitted with 7.92 mm MG 15 machine-guns as 'emergency armament'. For the time being, availability of the 7.92 mm MG 81 machine-gun intended in the first instance could not be guaranteed. As a consequence, all possible armament installations using the MG 15 and MG 131 had to be investigated first.

On 2 June 1937, less than five weeks after the death of *Generalleutnant* Wever, the Ernst Heinkel Flugzeugwerke was given the go-ahead to begin construction of a full-scale mock-up of its Project P 1041 proposal. Just short of three weeks later, on 22 June, the company submitted to the RLM the following timetable:

- Initial inspection of mock-up: 1 July 1937
- Final inspection of mock-up: 1 August 1937
- First prototype cleared for flight: 1 June 1938
- Delivery of first prototype: 1 September 1938
- Delivery of first pre-production series a/c: 1 October 1938

On 6 August 1937 the visual mock-up of Heinkel's response to the 'Bomber A' specification was inspected at the company's Rostock-Marienehe factory by experts from Branch LC II of the *Technische Amt* and *E-Stelle* Rechlin. Their joint verdict was as follows:

- Vision and firing angles satisfactory
- Crew space conditions insufficient
- Instrument panel layout in the too-narrow fuselage deficient

Final inspection of the suitably modified mock-up took place on 5 November 1937, three months later than originally planned. On that day, the RLM allocated Project P 1041 an official designation: He 177. Following the inspection, authorisation was given to start preliminary construction work on the new bomber, with an instruction to plan the aircraft for three operational ranges:

- Equipment Condition 'A' (short-range bomber): 2,000 km (1,243 miles)
- Equipment Condition 'B' (medium-range bomber): 3,000 km (1,864 miles)
- Equipment Condition 'C' (long-range bomber): 5,000 km (3,107 miles)

During a conversation with Prof Dr-Ing Ernst Heinkel immediately following the final inspection, Ernst Udet unexpectedly mentioned that the He 177 was no longer needed. After a short pause, Udet went on to express his belief that an air war against Great Britain was out of the question, and that for the coming conflicts in mainland Europe Hitler and Göring advocated the use of only twin-engined medium bombers capable of diving attacks. Despite this, Udet announced, for the time being the He 177 could be developed further as a 'research design'. He also thought conceivable that the He 177 could be planned for the *Kriegsmarine* (German Navy) as a long-range maritime reconnaissance-bomber aircraft. But that kind of aircraft too had to be able to dive to conform to the short-sighted ideas of the OKL.

Prof Dr-Ing Heinkel replied that such a big, four-engined aircraft would never be capable of

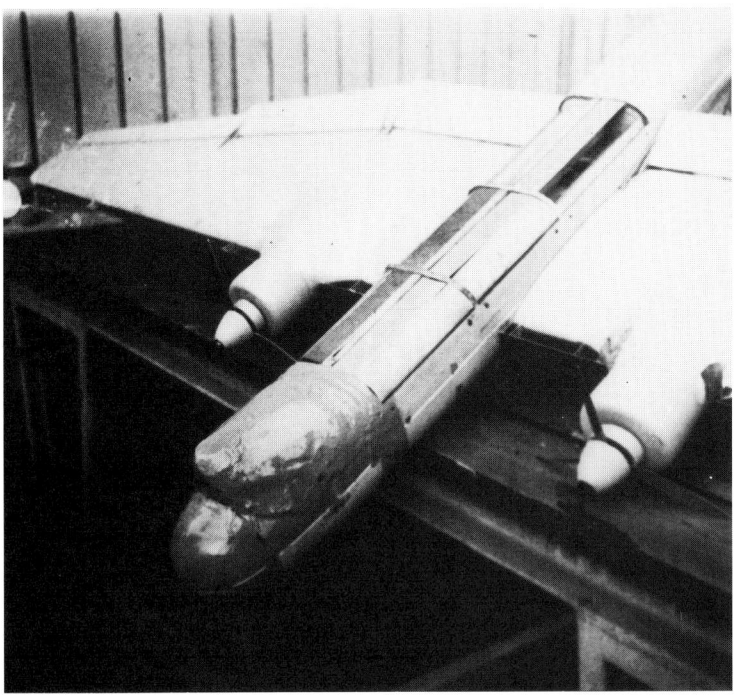

Two views of an early He 177 wind tunnel model.

diving attacks. But Udet, comparing the He 177 to the twin-engined Ju 88, then being tested in prototype form, could see no basic structural problems to prevent the Heinkel bomber from diving, as long as the airframe was strengthened accordingly. The inevitable increase in all-up weight should, he suggested, be parried by two twin-wheeled main undercarriage units fitted behind the DB 606 coupled engines.

In early 1938 the RLM made further demands, insisting on a range of 6,700 km (4,163 miles) at an altitude of 4,000-6,000 m (13,123-19,685 ft), with the crew (now increased to four) to be accommodated in a fully-glazed cabin. Furthermore, Heinkel's design office had to take steps to ensure that the aircraft, equipped with dual controls, would have 'good-natured' handling characteristics, so that it could be flown by pilots of average ability.

In August 1938, during a conference at Karinhall (General Göring's country residence north of Berlin), General Hellmuth Felmy, commander of *Luftflotte* (Air Fleet) 2, was asked by Göring to clarify the operational potential of the *Luftwaffe* in the event that *Fall Blau* (Case Blue: the codename given to plans for air warfare over and against Great Britain) was implemented. A few days later Felmy had the unenviable task of establishing that

> 'With the means presently available, we can only count on harassing effect. Whether this would lead to a wearing down of the British will to fight depends in part on imponderable factors. A war of extermination against England seems out of the question.'

These were the fundamental points of his comprehensive analysis, which was based on the fact that in normal circumstances a twin-engined medium bomber could only carry at most 500 kg (1,102 lb) of bombs over a distance of 700 km (435 miles) – figures based on the assumption that the whole of Western Europe had already been occupied by German forces to provide suitable operational bases for the *Luftwaffe*'s bomber formations. Given that the He 177 could hardly be brought up to operational status during the next two years, Göring responded by allocating the highest development and production priority to the Ju 88.

The origins of the Ju 88 lay in a 1934 RLM specification for a heavily-armed *Kampfzerstörer* (Heavy Fighter/Bomber-Destroyer) capable of operating as a bomber, bomber-destroyer, reconnaissance and close-support aircraft. This requirement was superseded in August 1935 by an RLM specification for a *Schnellbomber* (Fast Bomber): a three-seat, twin-engined bomber with minimal defensive armament, a bomb-load of 800-1,000 kg (1,764-2,205 lb) and a maximum speed of 500 km/h (311 mph), which could be built in no more than 30,000 man-hours per aircraft. Henschel (Hs 127), Messerschmitt (Bf 162) and Junkers responded, with the latter company's Ju 88 (single fin/rudder) being selected over the Ju 85 (twin fin/rudder) and the other competitors. Ironically, the design of what was to become the most versatile of all German bombers was entrusted to W.H. Evers and A. Glassner – the latter a US citizen and both having gained considerable experience in the art of modern stressed-skin construction while working in the US aircraft industry!

The Ju 88 V1 (D-AQEN/Wk-Nr 4941), the first of close to 14,700 Ju 88s to be built, conducted its maiden flight on 21 December 1936. The initial DB 600a 12-cylinder liquid-cooled radial engines gave way on later prototypes to the superior Jumo 211, trials having shown that the latter powerplant offered a maximum speed of 520 km/h (323 mph) compared to 465 km/h (290 mph) for the DB 600a-powered aircraft.

When the decision to implement large-scale production of the Ju 88 was taken, there were only three prototypes on hand. Then, following instructions from the *Technische Amt* of the RLM, this light *Schnellbomber* had to be redesigned as a heavy, strongly-armed dive-bomber.

In all, 10 Ju 88 prototypes were built, with the most obvious changes (other than the powerplant switch) being the introduction of a fourth crewman, a multi-panel glazed nose for the bombardier, a ventral cupola fitted with an aft-firing 7.92 mm MG 15 machine-gun operated by the radio operator, and slatted dive brakes outboard of the engine nacelles in strengthened wings; changes necessary if the Ju 88 was to be used successfully as a dive-bomber.

Development and testing of the Ju 88 continued throughout 1938, against a backdrop of political turmoil in Europe caused by Hitler's increasingly aggressive and expansionist foreign policy. His first success had been in regaining control of Saarland, placed under international administration by the Treaty of Versailles in 1919 (with France allowed to exploit the huge coal reserves) but recovered by plebiscite in January 1935. The following year German troops marched into Rhineland, the heart of German heavy industry, occupied by Allied troops from 1919-30 as part of the Treaty. The annexation was a huge gamble, but there was no opposition from the likes of Great Britain and France.

That same year, Hitler began to pressure Austria's government for unification. The pressure grew until, two years later, in March 1938, German troops marched over the border and onto Austrian soil. Shortly thereafter, the *Anschluss* (Union) between Austria and Germany was announced, with Austria now seen as the Ostmark of the 'Greater German Reich'.

Next came the turn of Sudetenland, an area in Bohemia awarded to Czechoslovakia in September 1919 by the Treaty of Saint Germain-en-Laye between Austria and the victorious Allied powers. Pressure was brought to bear by Germany itself and the German-speaking minority in Sudetenland in a campaign that grew in impetus following the *Anschluss* between Austria and Germany; and once again a policy of appeasement by Great Britain and France allowed Hitler to achieve his goal in October 1938 with the promise that it would be his 'last territorial demand in Europe'.

On 24 October 1938, after the Sudetenland crisis had been resolved to Germany's satisfaction, Göring and Milch held another discussion regarding *Fall Blau*: air warfare against Great Britain. In short, the first such air raid was to be carried out by all available *Luftwaffe* bombers, including those assigned to training units. Just two days later, another *Fall Blau* conference took place at Karinhall, during which *Oberst* Jeschonnek suggested that '. . . as many He 177s as possible should be authorised for the *Luftwaffe*; at the very least four *Geschwader*.' This proposal was based on the belief that such an attack could not be carried out before autumn 1942; only then, according to Jeschonnek, could about 500 He 177s be ready for operational use.

Hitler's and Göring's plans for 'an immediate five-fold increase in *Luftwaffe* potential' gave rise to incredulous astonishment among the RLM experts. The *Technische Amt* objected on account of the expected material shortages. Apart from that, the available fuel reserves would not permit such a large-scale expansion for the time being: to operate an air force of such a size would require 80 per cent of the world's production of aviation fuel! In the end Göring, who was often apt to act without the help of specialised knowledge, had to agree to a reduced aircraft production programme. Nevertheless, the principal items remained as before: the Ju 88, hardly-tested but which Göring already believed to be a 'high-performance dive- and inclined glide-bomber'; and the He 177, which existed only in mock-up form. Thus, wishful thinking at the highest level effectively determined the course of Germany's air armament policy as the threat of all-out war in Europe grew.

On 15 March 1939, less than six months after the occupation of Sudetenland, German troops marched into Prague and took over Bohemia and Moravia. In truth, the annexation of Sudetenland in September 1938 had merely been a preliminary move to enable the execution in full of *Fall Grün* (Case Green: Hitler's plan to conquer Czechoslovakia). Much of the country's industrial and economic wealth, including an impressive arms industry, now passed into German control. The Czech Army, impressive in size and capability, was disbanded and much of its equipment, notably several hundred modern tanks, transferred to German Panzer divisions.

One week after *Fall Grün* was implemented in full, the port and hinterland of Memel (now Klaìpeda) in Lithuania, part of East Prussia before the First World War but lost as a result of the Treaty of Versailles, was also recovered by Germany.

Following the march into Czechoslovakia and the occupation of Memelland, endless discussions began concerning the earliest possible commencement of hostilities against Great Britain. A study prepared by the Operations Branch of the *Luftwaffengeneralstab* entitled Operational Objectives for the *Luftwaffe* in case of War against England in 1939, could not have painted a more negative picture for General Göring:

> 'Equipment, crew readiness and strength of Luftflotte 2 could not bring about decisive results within a short period of time in a war against England.'

Erhard Milch, promoted to *Generaloberst* but whose deteriorating relationship with Göring led to his being replaced as *Generalluftzeugmeister* by Ernst Udet in February 1938, had pointed out for quite some time that to commit the *Luftwaffe* to a war of aggression with the He 111 and hardly-tested bombers, such as the Ju 88, would be like stepping into an abyss. But Göring no longer seemed to be interested in the whole business, even though shortages in material supplies were becoming ever more acute: according to the experts, supplies of aviation fuel would only last for six months if no new sources of raw material became available. In addition, the stocks of drop-loads comprised relatively few SC 50, SC 250 and some SC 500 bombs; larger-calibre bombs existed mostly just as design drawings.

As 1939 progressed, Göring seemed only slightly concerned that production of the much-anticipated 5,000 Ju 88s would not be completed until April 1943. Hitler and Mussolini, whose forces invaded Albania on 7 April 1939, signed a military alliance (the so-called Pact of Steel) on 22 May 1939, which ostensibly saw the two governments agree not to initiate war until at least 1942. In fact, within 24 hours of signing the agreement Hitler had issued orders for war that same year, with Poland, effectively surrounded by German forces on three sides, the principal target. All the signs indicated a coming storm. As Hitler put it:

> 'The settling of our frontiers is a matter of military importance. We must enlarge our living space in the east . . .'

Y-Day, the date for the invasion of Poland, was originally 1 September 1939; but this was brought forward to 26 August following the signing of the German-Soviet Non-Aggression Pact on 23 August, then put back in the final hours of 25 August following the signing in London of the Anglo-Polish Treaty of Mutual Assistance. A flurry of diplomatic activity in the final days of August came to nought, and in the early hours of 1 September German forces, spearheaded by an initial attack by Ju 87B-1 dive-bombers, put *Fall Weiss* (Case White: the military invasion and destruction of Poland) into operation. Demands by Great Britain and France for the immediate cessation of hostilities went unheeded, and on 3 September both countries declared war on Germany. The Second World War had begun.

Although the Ju 87 was the trump card in the German forces' *Blitzkrieg* on Poland, two of the new generation of twin-engined medium bombers, the Do 17 and He 111, made a significant contribution to the campaign which was all but over within the month. Certainly the role of the *Luftwaffe* was pivotal, as General Kesselring, commander of *Luftflotte* 4 during the campaign, acknowledged later:

> 'The Polish campaign was the touchstone of the potentialities of the Luftwaffe and an apprenticeship of special significance. In this campaign the Luftwaffe learned many lessons . . . and prepared itself for a second, more strenuous and decisive clash of arms.'

Despite the *Luftwaffe*'s success, 285 of the 1,935 aircraft committed to *Fall Weiss* were lost, just over one-third of which were twin-engined bombers, and approximately half of its existing stocks of bombs were used. As a consequence, under Milch's resolute direction, the *Luftwaffe*'s bomb stocks were continuously built up and all prerequisites for the coming greater conflict were created. As for the Ju 88, the first A-0 pre-production series examples had been assigned to *Erprobungskommando* (EKdo: Test & Evaluation Detachment) 88 in mid-1939, which in turn formed I/KG 25 in August of that year, only to be redesignated I/KG 30 'Adler' (Eagle) on 22 September 1939.

However, the decisive battle would come in the skies over England, and the lack of four-engined strategic bombers for the *Luftwaffe* would exert a considerable and lasting influence on the course of the Second World War. At that time, nobody had any idea when to expect the He 177.

Slow Progress

The first serious delays to the He 177 project occurred early in 1938, during the construction of the first experimental prototype. The RLM had left it too late to initiate successful manufacture of the Daimler-Benz DB 606 experimental engine (actually two DB 601 12-cylinder liquid-cooled inline engines coupled side-by-side); and nothing had been done about the promised DB 606 dummy installation units. The pressure to find a solution increased when, during a visit to the RLM by Dipl-Ing Hertel on 28 September 1938, an option was taken on six He 177 prototypes, all of which had to be developed and built with the utmost speed.

In an urgent letter dated 1 October 1938, Dipl-Ing Hertel practically begged *Fliegerhauptstabs-Ing* Eissenlohr, Chief of the LC 3 (Powerplants) section of the *Technische Amt*, to force Daimler-Benz to deliver DB 606s. But only two prototype engines were available, the DB 606 V5 and V6, both of which had originally been kept in readiness to power the private-venture He 119; a high-speed unarmed reconnaissance-bomber powered by a DB 606 coupled powerplant housed in an aerodynamically very clean fuselage to drive a four-bladed propeller via a long transmission shaft fitted inside the glazed nose section. Designed by Siegfried and Walter Günter amidst great secrecy in the mid-1930s, the first of eight prototypes of this highly-advanced aircraft flew in summer 1937. On 22 November 1937 the He 119 V4 established a world speed record for its class of 505 km/h (313.79 mph) over a 1,000 km (621.37 miles) closed circuit whilst carrying a 1,000 kg (2,204.62 lb) payload.

Despite the record-setting success of the V4, the He 119 was not accepted for series production for the *Luftwaffe*. Nevertheless the engine delivery situation in general, and for the He 177 in particular, remained completely unsatisfactory.

Apart from the lack of engines and the on-going search for optimum horizontal tail surface and fin/rudder assembly forms in the *Deutsche Versuchsanstalt für Luftfahrt* (DVL: German Aviation Experimental Establishment) wind tunnel, it was essential first to thoroughly research the vibration safety factor.

On 12 November 1938, even though all of the problems had by no means been solved in detail as yet, the RLM issued an official preliminary notification (LC 7 No 1661/38 geh) for the construction of the first, unarmed prototype, the He 177 V1. On the strength of that notification, the delivery date of the V1 with DB 606 coupled engines, and at an agreed cost of RM1,357,000, was confirmed as 27 February 1940. A short while later there followed an option for the manufacture of a total of six He 177 prototypes (V1-V6), all of which had to be capable of diving attacks. Once again, the use of the desired DB 606 engine installation met with considerable scepticism on the part of Prof Dr-Ing Heinkel and his closest colleagues. However, a proposal to the *Luftwaffengeneralstab* to fit two of the six prototypes with four separate Junkers Jumo 211 engines was rejected on the grounds that they lacked diving capability.

An interim (Heinkel) works conference regarding powerplants took place on 17 November 1938, and it was agreed to fit only the He 177 V1 and V2 with DB 606 coupled engines to begin with; while owing to the powerplant situation the He 177 V3 and V4 were each to be fitted with four separate engines.

In January 1939 the RLM received Structural Description No 617 covering the He 177 heavy bomber. In the meantime, on account of structural strengthening and a more extensive equipment fit, the He 177's take-off weight had risen from 25,000 kg (55,115 lb) to 29,000 kg (63,934 lb). This increase in weight, the main causes of which were additional equipment and the need to use frontal engine radiators large enough to provide sufficient cooling instead of smaller orthodox radiators augmented by surface evaporation cooling, was the reason why the originally estimated operational speed of 500 km/h (311 mph) had been reduced to 460 km/h (286 mph).

As the problems with the DB 606 engine seemed endless, Prof Dr-Ing Heinkel and Dipl-Ing Hertel initiated discussions with the Junkers Flugzeug- und Motorenwerke (JFM) with a view to combining two Jumo 211s to produce a Jumo 212 coupled engine. Performance estimates had shown that the Jumo 212, though smaller in size, would offer the potential of an unexpected increase in speed of possibly 15 km/h (9 mph). One year later, another possibility appeared when the combination of two Jumo 213s instead of two 211s came under consideration. As a result of these options, on 24 February 1939 the number of He 177 prototypes was doubled to 12 aircraft, to enable various combinations of powerplants to be tested.

By April 1939 it was still not quite clear exactly when the 52 DB 606 engines initially requested (for both the He 119 and He 177) would finally become available. By then, without any consideration of the need for reserve engines, the RLM had ordered only 42 DB 606s from Daimler-Benz. Although nobody yet had any idea if the DB 606 would go into series production, during his meetings with Prof Dr-Ing Heinkel in April 1939 General Udet indicated that the He 177 V1 must be ready for flight-testing without fail by 30

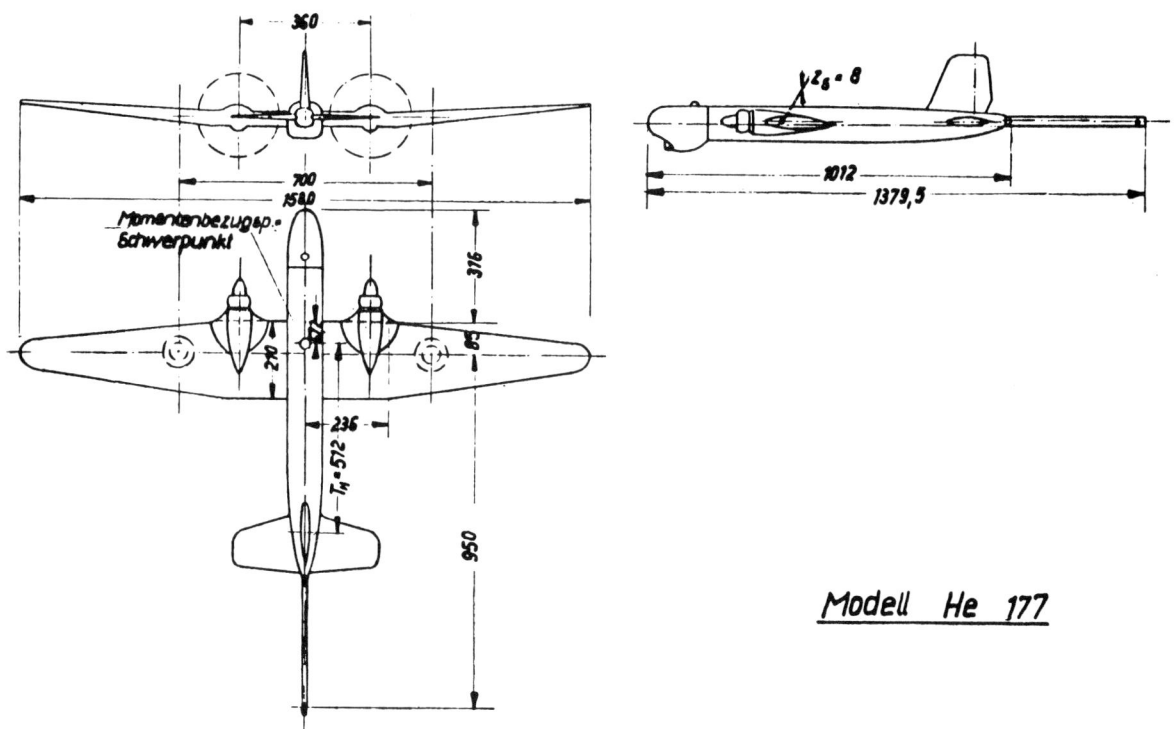

Three-view drawing of the He 177 V1 model.

August 1939. As for the DB 606, Udet agreed that a further six He 177s should be fitted with air-cooled double-row radial engines, of which the first should start its flight tests by the end of April 1940. An exchange of views between Heinkel and Udet resulted in the following timetable:

- He 177 with DB 606s: end of August 1939
- He 177 with Jumo 212s: end of January 1940
- He 177 with BMW 802s: mid-April 1940
- He 177 with surface
  evaporation cooling: mid-June 1940

At the same time, there were fierce arguments for and against the use of air-cooled double-row radial engines on the He 177. Apart from the BMW 802 (two coupled BMW 801 radials), which was not expected to go into large-scale production for another 2fi years, the RLM could not see any other ways of having a sufficient number of such engines produced on time. Already early in 1939, the demand for BMW 801s for the Ju 88 and Ju 90 could hardly be calculated.

The RLM therefore looked upon the Jumo 212 as the substitute solution for the DB 606. Consequently, when General Udet paid his next visit to the Rostock-Marienehe works on 26 April 1939, there ensued another discussion about the He 177 fitted with four separate engines and the soon-expected realisation of an aircraft fitted with a surface evaporation cooling system. In addition, members of the RLM commission were briefed about the on-going breaking load tests, work on the Fowler flap wing model, and the new cockpit mock-up.

Dipl-Ing *Leutnant* Carl Francke, chief of the *E-Stelle* Rechlin E-2 flight test section, was somewhat critical of the cockpit as originally planned (and similar to the one subsequently used on the He 177 V1), observing that it provided insufficient space and vision for the crew. For the time being, due to the very tight production timescale and because according to his Development Department the complete armament installation would only be fitted from the He 177 V4 onwards, Prof Dr-Ing Heinkel still planned to complete the V1 with the early, unarmed cockpit.

In the meantime, problems with the DB 606 escalated. Serious difficulties were encountered during the yet-to-be-finished parts manufacture at Daimler-Benz's Friedrichshafen gear wheel factory. This set back the start of DB 606 series production by a further three months, which in turn even cast doubt on the 1940 delivery schedules. Matters were just as critical regarding the completion of even a few DB 606 prototype units in summer 1939 to power at least the first few He 177 prototypes.

Cleared to Fly

Despite the verbal promises of the RLM in 1939, there was only one binding contract for the first He 177 prototype and an option for a further eight. All six previously-requested pattern aircraft with air-cooled engines as well as two aircraft with surface evaporation cooling, had in the meantime become victims of the RLM's dive-bomber mania. All that mattered to the *Technische Amt* was whether the He 177 V1 should be followed by eight or 12 additional prototypes powered by coupled engines. Of these aircraft the V1 was to be unarmed, the V2 and V3 would each be fitted with a simplified bomb-release mechanism, and only the V4 would feature a complete military equipment fit as well as the afore-mentioned MG 15 machine-gun 'emergency armament'.

Quite unexpectedly, on 20 September 1939 the He 177A-0 pre-production series was authorised for production with the considerably improved 'Cabin 3' and a new bomb-release mechanism. Also, instead of the initially-demanded catapult launch installation the aircraft were now to be accelerated to take-off speed by two 2,500 kp (551 lb st) or four 1,300 kp (2,866lb st) rocket boosters.

During summer 1939 production of the component parts and assembly of the He 177 V1 was underway, and by late October even the first flight seemed not far off. Engine ground-running tests had revealed no significant problems. On 23 October Dipl-Ing *Leutnant* Francke was delegated to Rostock-Marienehe to familiarise himself completely with the He 177 V1. In the meantime, General Udet had arranged with Prof Dr-Ing Heinkel to pay Francke a 'Pilot's fee' of RM10,000 for flight-testing the aircraft.

On 1 November, during another visit to Rostock-Marienehe, General Udet expressed the opinion that the *Luftwaffe* command considered a fast dive-bomber completely sufficient to deal the Royal Navy destructive blows in its ports, and thus force Great Britain to surrender. It was only after Prof Dr-Ing Heinkel's question as to what would happen if the Royal Navy's Home Fleet moved out of the range of *Luftwaffe* Ju 88s, that Udet conceded and agreed to again allocate a higher priority to the bigger He 177.

Udet's decision was perforce the basis of the first large-scale production series contract for 800

The He 177 V1 (CB+RP/Wk-Nr 00 0001), flown for the first time on 9 November 1939.

He 177s, delivery of which had to be completed by the end of April 1943. Simultaneously, the RLM increased the number of pre-production series He 177A-0s on order to 30. A little later, Udet wanted to increase the forthcoming He 177 production programme to no less than 120 aircraft per month by summer 1940! Even if this bit of wishful thinking by an ex-First World War fighter pilot was totally removed from reality, his change of mind concerning the priority given to the He 177 did finally lead to more open-mindedness and receptivity regarding long-range heavy bombers.

Heinkel took advantage of this favourable atmosphere to postpone the He 177 V1 expected first flight date until November 1939, there still being numerous technical problems with the aircraft's rudder installation and material supplies for the undercarriage units.

Despite all this, the He 177 V1 (CB+RP/Wk-Nr 00 0001) took off on its first flight earlier than expected, on 9 November 1939. Piloted by Dipl-Ing *Leutnant* Francke, the aircraft made a 20-minute flight, during which the undercarriage remained locked down. An altitude of 2,000 m (6,562 ft) was reached at an all-up weight of 16,000 kg (35,274 lb), the take-off and climb being described by Francke as 'very good'. Forces acting on the rudder were 'too high', but the aileron forces he judged to be excellent, and the same applied to stability around the normal axis. But there were also negative aspects: the shock-absorbing qualities of the undercarriage units, their brakes, poor cabin ventilation – and the powerplant installations. Then, after the oil temperature climbed to 120°C (248°F), Francke decided to terminate the flight prematurely.

The second test flight took place on 20 November and lasted for 14 minutes. Apart from the very slow retraction of the two twin-wheeled main undercarriage units and a minor change in trim while retracting the Fowler flaps, there was a sudden oscillation caused by one defective air brake flap. On the other hand, the various faults reported after the first flight had already been partly eliminated.

On the third test flight, Francke was accompanied by test engineer Naumann and flight engineer Wüppelmann. At first, things proceeded smoothly; but on reaching an altitude of just 200 m (656 ft) the He 177 V1 suddenly began to vibrate after one of the main undercarriage doors came loose and bent itself beneath the wing.

A further seven test flights were accomplished from 6-18 December, all except one (when the V1 was taken aloft by Heinkel works pilot Rieckert) flown by Francke. On the fourth test flight Francke had noticed excessive stick forces, and despite the enlarged elevators in-flight stability was still found to be insufficient. The sixth test flight had to be terminated prematurely due to a substantial loss of oil; and during the next flight, while gliding at about 400 km/h (248 mph), the entire airframe was suddenly shaken by an odd vibration. Shortly afterwards, on 9 December, the eighth test flight also had to be terminated early, this time because of defective undercarriage indicators. Two days later, the V1 was grounded following damage to its elevator assembly and did not start rudder test flights until mid-December.

The He 177 V1 flown by Heinkel works pilot Schäfer, seen on final approach to Rostock-Marienehe. Note the twin-wheel main undercarriage units.

After flight-testing resumed, Francke faced a more serious fault on 17 December, when he had to return to base after a complete failure of the electrical system aboard the 16,960 kg (37,390 lb) prototype.

From February 1940 onwards the flight-test programme embraced propeller measurements and evaluation of the DB 606 coupled engine installation, as well as improvements made to the ailerons by *E-Stelle* Rechlin personnel. Simultaneously, the DVL carried out a series of pressure distribution measurements and ditching tests with a 1:14 scale model of the He 177.

On 26 February 1940, and following on from the preliminary notification of 12 November 1938, the State Minister of Aviation finally issued the

The manually-operated tail MG 131 machine-gun on the He 177 V2.

THE HE 177 IS BORN   19

The He 177 V2 (CB+RQ/Wk-Nr 00 0002), photographed at Rostock-Marienehe during a break in its flight-test programme.

official contract for the construction of the He 177 V2-V6 at a total cost of RM6,201,750. A little later came an order for the He 177 V7 and V8, in accordance with Heinkel's offer of 9 February 1940.

In spring 1940 the He 177 V2 (CB+RQ/Wk-Nr 00 0002) and V3 (D-AGIG/Wk-Nr 00 0003) conducted their maiden flights at Rostock-Marienehe. The V2 differed from the V1 in having an enlarged fin/rudder assembly with aerodynamic balance, while the V3 featured an improved tailplane adjustment pinion.

The V3's short life came to an end on 24 April 1940 when, due to faulty trimming, the aircraft crashed at Geldorf during an inclined flight test, killing the crew captained by Heinkel works pilot Rieckert. Two months later, on 27 June, the V2 was lost while conducting minimum permissible speed trials. Due to a wrongly selected propeller pitch setting, the aircraft ditched in the Baltic Sea north of Graal/Müritz from a height of just 400 m (1,312 ft). The four-man Heinkel works test crew onboard lost their lives. As a result of these crashes, only the He 177 V5 (PM+OD/Wk-Nr 00 0005) remained to continue flight-testing.

These accidents were the subject of much discussion during a meeting between Prof Dr-Ing Heinkel, director Hayn and Heinkel's Technical Director, Dipl-Ing Robert Lüsser. To conclude the meeting, Dipl-Ing *Leutnant* Francke presented his view of the forthcoming He 177 development stages as follows:

'The 'Bomber B'* has evolved into a much larger aircraft. The 'Bomber B' I really had in mind was a [13.6 tonnes] 15-ton or so replacement for the Ju 88.

An early view of the He 177 V2, taken prior to application of the four-letter radio call-sign to its fuselage.

Inspection of the He 177 V3 (D-AGIG/Wk-Nr 00 0003) at Rostock-Marienehe. In the background are several He 111 bombers ready for collection.

The He 177 V2's life was relatively short, the aircraft crashing north of Graal/Müritz on 27 June 1940.

The He 177 V5 (PM+OD/Wk-Nr 00 0005) carried only 'emergency armament'. This photograph shows the aircraft during take-off distance determination trials at Rostock-Marienehe.

Other companies want to build it weighing around [20 tonnes] 22 tons, but I can already visualise it being [22.6 tonnes] 25 tons, and in that case I would rather prefer the He 177.'

On 28 June 1940 Prof Dr-Ing Heinkel had an animated discussion in Berlin with General-Ing Lucht, who already envisaged the He 177 as the future standard long-range aircraft for maritime warfare. But only a few days before, General Udet had observed: 'There must surely be something basically wrong with that machine!'

His comment followed the loss of the He 177 V2 and V3 due to then still-unknown causes. Director Kloppenberg of JFM was quick to seize this opportunity, seeing it as a last chance to stop the He 177 series production. General-Ing Lucht was later to quote the words of Kloppenberg: 'I am going to kill the [Do] 217 and the He 177 before they are born!'

'He was intent on building the Ju 288 before it was ready!' added Lucht.

In December 1940 the He 177 V1 was fitted with a revised fin/rudder assembly. After flight-testing, this aircraft was to serve as a reserve with the manufacturer and carry out various minor tests. At that time the V5 was also in for repair at Heinkel's Rostock-Marienehe works: while landing, the 23,000 kg (50,706 lb) aircraft's undercarriage brakes had locked and broken the engine bearer blocks. The Heinkel works' He 177 test programme included improvement of the Fowler flap position as well as flights with a take-off weight up to 29,000 kg (63,934 lb). At that stage only the He 177 V4 (Wk-Nr 00 0004) was on hand to continue the flight tests, the aircraft being used mainly to speed up aileron trials. For the intended inclined flight tests the V4 was equipped with ejector seats for the crew and a large brake parachute. The Heinkel Development Department expected these tests to start in January 1941.

The He 177 V6 (BC+BP/Wk-Nr 00 0006) was completed in early December 1940 and made its first flight on 18 December. Following flight tests, the main aspects of its test programme comprised trials involving the bomb-bay doors and the

---

* A specification issued by the RLM in July 1939 to Arado (Ar 340), Dornier (Do 317), Focke-Wulf (Fw 191) and Junkers (Ju 288) for a twin-engined high-altitude medium bomber to replace the Ju 88 and He 111. Features were to include a pressurised cockpit for the crew of three or four, remote-controlled gun barbettes, a range of 3,600 km (2,237 miles) and a 2,000 kg (4,409 lb) bomb-load. The Fw 191 and Ju 288 were selected for further development, but the entire programme was abandoned in June 1943, due primarily to shortages of strategic materials.

The He 177 V4 (Wk-Nr 00 0004) during its take-off run and on final approach to landing. This prototype was primarily used to gather performance data and for inclined flight trials.

operation of defensive weapon positions. Later, the aircraft was retrofitted with a 20 mm MG FF cannon in the lower fuselage nose section. There then followed the addition of a larger fin/rudder assembly and the exchange of prototype engines for DB 606A-1/B-1 (A-1 to port, B-1 to starboard) pre-production series units.

From about February 1941 onwards the V6 began to undergo operational trials. The same applied to the He 177 V7 (SF+TB/Wk-Nr 00 0007), which was completed early in 1941 and, after initial flight tests at Rostock-Marienehe, was used for a while for automatic flight control trials. Upon their completion, the V7 was also transferred to operational trials.

Unfortunately, new delays had to be accepted in the area of powerplant development when several components in recently-delivered DB 606 engine units had to be returned to Daimler-Benz. The lack of operational reliability of the DB 606 V7-V11 in particular gave rise to a voluminous correspondence between the RLM, Heinkel and

The He 177 V6 (BC+BP/Wk-Nr 00 0006) takes off at the start of another flight test.

Daimler-Benz, but did not lead to any fundamental improvement of the situation.

Type Trials and the He 177A-0

Early in 1941, the entire He 177 test programme rested on just three prototypes (V1, V5 and V6) and the first pre-production series aircraft, the He 177A-01 (DL+AP/Wk-Nr 00 0016). After being fitted with improved engines, the V5 resumed longer-duration automatic flight control trials. At that time the He 177 V6 had just been delivered to *E-Stelle* Rechlin to be fitted with the flexible armament required for the planned operational trials. The He 177 V7, which had been armed from the start, was simultaneously undergoing armament tests at *E-Stelle* Tarnewitz, these having begun in February 1941.

The He 177A-01 featured the extensively reworked production series cabin as well as the new tail assembly. However, as the required weapon installations had not arrived on time, the aircraft had to be test-flown unarmed. Next to checking the flying characteristics and some speed-measuring flights, it was especially urgent to test the bomb-bay installation.

The second pre-production aircraft, the He 177A-02 (DL+AQ/Wk-Nr 00 0017)), was cleared for flying the following month. The number of He 177A-0 pre-production series aircraft built by Heinkel at Rostock-Marienehe and Oranienburg and by Arado Flugzeugwerke at Brandenburg-Neuendorf finally amounted to 35:

- He 177A-0 Wk-Nr 00 0016-0035 (Rostock-Marienehe)
- He 177A-0 Wk-Nr 32001-15 (Oranienburg)
- He 177A-0 Wk-Nr 335001-005 (Brandenburg-Neuendorf)

The first of these aircraft arrived for trials at *E-Stelle* Rechlin on 19 May 1941. However, there were repeated delays because of frequent engine failures and faults. The fitting-out of 'front-line' He 177s initially took place at Rechlin-Lärz, but from summer 1941 onwards this work was undertaken at Lüneburg. In addition, Lüneburg became a staging point for the selective instruction of pilots and crew members on the He 177, although in summer 1941 only the He 177 V7, V8 (SF+TC/Wk-Nr 00 0008) and A-02 were available for this task.

Following comprehensive loading trials, the V8 was flown to *E-Stelle* Tarnewitz to test the accuracy of the bomb-sights. On 8 July 1941 at around

A fine study of the He 177 V7 (SF+TB/Wk-Nr 00 0007), the second pattern aircraft used for defensive armament trials. This view clearly shows the 20 mm MG FF cannon mounted in the lower nose position.

18:50 hours, the aircraft was supposed to fly back to Rostock-Marienehe, but during the take-off run it collided with a parked Ju 52/3m, which was almost completely wrecked.

Meanwhile, further work was underway at Lüneburg. Along with various modifications that had to be carried out on production series aircraft and several other minor tests, the endurance trials and instruction of additional pilots continued unabated. But all this was delayed by numerous engine defects; and matters were made worse by the limited number of airworthy He 177s available, with the He 177 V5 deployed at Brandis for a while and the V8 under repair at Rostock-Marienehe following its take-off mishap.

A few days later, on 14 August, the He 177 V6 was grounded at Bordeaux due to a cracked engine bearer. Conversely, the He 177 V7, demonstrated to *Reichsmarschall* Göring (promoted from General in June 1940) near Paris the following day, gave a very good show – apart from the landing. Major Peterson of KG 40, who was piloting the aircraft, touched down a little too late and performed an impressive ground loop. The *Reichsmarschall* and his escorts were already preparing to witness the inevitable undercarriage smash-up, but the aircraft overcame the 'pilot error' without any damage. Göring was elated and shouted delightedly: 'At last, an undercarriage that can take it!'

Despite every effort to make the He 177 fit for operations as soon as possible, the general situation was still highly unsatisfactory. Prof Dr-Ing Heinkel blamed this on still-outstanding parts of equipment on whose production he had no influence. Apart from the remote-controlled weapons there were bottlenecks with the Schwarz wooden propellers as well as the VDM duralumin propellers. At the Rostock-Marienehe works in early August there were only 12 wooden propellers in store for the He 177; not a single propeller, wooden or duralumin, was available at Arado's Brandenburg-Neuendorf licence production plant. The VDM plant near Frankfurt am Main could not deliver its propellers on time, either: until May 1941 not a single variable-pitch propeller had found its way to Rostock-Marienehe. It was not even possible to procure

Four of the first five He 177 prototypes were lost in crashes; here the remains of the V1 smoulder after its demise on 3 October 1941.

spare propellers for the He 177 V6-V8, then keeping the flight-test programme going. Additional delays remained the rule.

A new series of tests with the He 177 V1 began in September 1941, the object being to reduce the bomber's tendency to ground loop. In October, the Heinkel works completed the powerplant and tailplane assembly changes on the He 177 V5, which then became available once again for flight tests. Earlier that same month, after extensive overhaul, the He 177 V8 was flown back to *E-Stelle* Rechlin.

For these reasons, it was primarily the He 177 V6 that was available for training and instructional purposes at Lager-Lechfeld, with predictable results: by late summer 1941 the well-used airframe had begun to show the effects of continuous stress-fatigue loading tests. While at Lager-Lechfeld – also home at the time to the He 177 V7 and the He 177A-0/V15 (Wk-Nr 335001; the first Arado-built He 177) – the V6 was flown mainly by pilots of KG 40, who assessed the He 177's flying characteristics as well as the take-off and landing as 'very good'.

Meanwhile, the He 177 V1 was severely damaged during a heavy landing at Rostock-Marienehe on 3 October 1941. The main undercarriage units were sheared off due to excessive lateral forces, but the crew, captained by Heinkel factory pilot Schuck, escaped uninjured. By the time of the accident the He 177A-02 and A-03 (DL+AR/Wk-Nr 00 0018) were flying at *E-Stelle* Rechlin, both having arrived there in October 1941.

Apart from the late-autumn weather, it was once again the defective engines that made it all but impossible to adhere to any flight-testing timetable. Prof Dr-Ing Heinkel's sharp criticism of the DB 606 powerplant did not go unheeded by Daimler-Benz, but in an official communication by the Chairman of the board of directors, Dr-Ing W. Kissel and Director Nallinger, Heinkel's criticism of the DB 606 was glossed over and rendered harmless. The engine's lack of operational reliability was explained away by the fact that the units in question were only DB 601A/E engines that had been provisionally rebuilt to give DB 610E power rating, and were thus considered to be underdeveloped.

Despite that, the life expectancy and operational reliability of the DB 606A-0-series powerplants (coupled DB 610E engines) already delivered proved to be hardly any better. In any case, according to Daimler-Benz the frequent powerplant troubles were actually caused by the coolant installation which provided irregular water circulation; while blame put on the structure of the DB 606 itself was considered misleading. There was some truth to this: the He 177 V8 and A-02 were the first examples of the bomber to be fitted with new, enlarged engine cowlings, and it was only after this modification that the DB 606 enjoyed a more sustained service life.

Left and opposite (top): Although fitted with the original, squarer rudder, the He 177 V8 (SF+TC/Wk-Nr 00 0008) was also fitted with shortened Fowler flaps and new, enlarged engine cowlings and radiators.

In August 1941 the He 177 V7 was assigned to IV/KG 40 in France for trials and instruction purposes. The aircraft made a notable impression with the KG 40 aircrews.

In mid-November 1941 the He 177 V6 and V7 were flying as training aircraft for KG 40 crews, while the V5 was at last set to start the long-planned inclined flight tests on 12 November. As for the He 177A-0 pre-production series aircraft, a total of 11 (A-02 to A-07 and A-011 to A-015) were to be at the disposal of *E-Stelle* Rechlin. Apart from six aircraft that, among other things, were used for powerplant, flight characteristics and automatic flight control tests, there were another three He 177A-0 machines for the development of defensive and offensive (drop-load) trials. In addition, the He 177A-05 (GA+QN/Wk-Nr 00 0020), A-06 (GA+QO/Wk-Nr 00 0021), A-011 (GA+QT/Wk-Nr 00 0026) and A-012 (GA+QU/Wk-Nr 00 0027) were doing their share of flying at Rostock-Marienehe.

A special heavy-load crane was developed by Heinkel to salvage crash-landed large aircraft including the He 177, thus clearing runways and minimising disruption.

Suddenly, in January 1942, the RLM grounded all He 177s. A subcontractor to Daimler-Benz had inadvertently used aluminium instead of duralumin rivets on the joints between the reduction gears and propeller shafts. The whole flight test programme had to be put on hold until new propeller shafts were delivered.

The test programme had resumed for only a short while when, much to everyone's surprise, *Generalfeldmarschall* Milch, who had once again assumed the responsibilities of *Generalluftzeugmeister* in late 1941 following the suicide of General Udet, once again ordered the grounding of all He 177s; the reason given was several defects found in the propeller pitch-change mechanism, as well as several engine fires.

A new reversal of fortune affected the flight test programme in March 1942. Test pilot Peter and flight engineer Daub were detailed to conduct an inclined flight test on 24 March, but at an altitude of 5,000 m (16,404 ft) the engine revs suddenly fell off and the oil pressure dropped rapidly. After the instruments had again reached their normal values, the crew began an approach to Rostock-Marienehe for a careful landing. However, due to being almost completely dazed, Peter noticed too late that his approach was adrift of the main runway. An attempt to go around was brought to an abrupt end by the sudden failure of the starboard powerplant, and the aircraft crashed within the boundaries of the Heinkel works. After repairs this aircraft was the only He 177 prototype cleared for flying four months later.

For quicker salvage of crashed aircraft, Heinkel's Prague branch station submitted plans to *E-Stelle* Rechlin for a motorised crane weighing 27.2 tonnes (30 tons) that could lift a crashed He 177 or similar large aircraft with ease, and thus prevent long interruptions in flying operations on an airfield.

Engine failure also accounted for the loss from low altitude of the A-05 while conducting an oil cooler measurement test flight on 13 June 1942. The aircraft had taken-off at 11:48 hours in the morning, piloted by Ahlhorn who was accompanied by measurement engineers Jarmatz and Piacenza, and flight engineer Storr. After the failure of one engine the 31,000 kg (68,343 lb) aircraft quickly lost height – the He 177 was impossible to fly on one engine at an all-up weight in excess of 28,000 kg (61,729 lb) – and crashed soon afterwards with fatal consequences for all on board. The subsequent crash investigation revealed the actual cause to have been the failure due to overstress of the internally geared and riveted engine coupling casing.

It is almost certain that this crash was the reason for a report by *Oberstleutnant* Petersen of the *Technischen Amtes* to *Generalfeldmarschall* Milch, requesting court-martial proceedings against Prof Dr-Ing Heinkel to find the guilty parties responsible for the loss of the four-man crew and the aircraft. But nobody could or would determine with any degree of certainty if the real reason for the crash had been a genuine material defect, or if it was rather a matter of structural failure. For that reason, all He 177s were grounded again – for the fourth time since January 1942.

On 23 June Prof Dr-Ing Ernst Heinkel finally faced *Generalfeldmarschall* Milch and declared his intention to clear up at once all faults on the He 177 and, in the event of others, to immediately inform the RLM. Despite this the *Kommandeur der Erprobungsstellen* (KdE: Commander of Test Establishments) ordered via the Council of Aviation Industry that from June 1942 onwards, tests carried out by aircraft manufacturers should be bound by the directions of the KdE. More bureaucracy and an encroachment on the progress of the He 177 flight test programme were the inevitable results.

But there were also brighter moments during 1942. On 27 January the He 177A-07 (GA+QP/Wk-Nr 00 0022) successfully completed an overload test at Lärz. On the strength of that, a few days later *Hauptmann* Mons of *Erprobungsstaffel* (E-Staffel: Test/Proving Squadron) 177 received an order to conduct a 12-hour flight with the He 177A-07. The flight route took the 30,000 kg (66,139 lb) aircraft on a 4,700 km (2,920 mile) round-trip from *E-Stelle* Rechlin via Bordeaux, Warsaw and Munich. Another tactical test involved a comparison flight between the He 177A-06 and a captured Supermarine Spitfire fighter on 11 February 1942, which highlighted well the He 177's performance capabilities against single-seat enemy fighters.

During a conference on 1 June 1942, the Heinkel works management ordered categorically that there should be no more delays to the He 177 flight test programme. To complete the still-outstanding tactical evaluation, it was necessary to finish the diving tests (involving the

V5) as quickly as possible. The first test, on 14 April, had proved negative due to the failure of measuring instruments. The second test, nine days later, was also a failure: test pilot Peter reached a speed of only 705 km/h (438 mph). Afterwards, the aircraft was slightly damaged during an air raid and had to go into the workshops for repairs.

The third diving test also failed to achieve the expected results. Then, during the follow-up demonstration flight, the dorsal gunner's cupola came loose and damaged the elevators. A few moments later the rear bomb-bay door also detached itself and sliced through the recently-fitted new horizontal stabiliser.

Another in-flight accident followed only a short while later, on 16 July 1942. While carrying out an inclined flight test with the He 177A-013 (GA+QV/Wk-Nr 00 0028), the pilot, Scherling, pulled out of the dive too sharply over the Plauer lake near Rechlin, and a wing broke off the aircraft. Fortunately all but one of the crew managed to save themselves by parachute.

Following this accident the *Technische Amt* immediately ordered speed restrictions for all He 177s used in flying operations. Thus, the maximum permissible speed at altitudes up to 3,000 m (9,842 ft) was limited to 500 km/h (311 mph), falling to 400 km/h (248 mph) at altitudes in excess of 6,000 m (19,685 ft). At the time only the He 177 V5 was cleared for flying (the V9 remained grounded until the arrival of new shock absorbers from Rostock-Marienehe, while the V8 had already been hangared for 47 days, not one of the urgently-needed reworked coupling sleeves having been delivered). But use of the V5 in further diving tests was put on hold due to low cloud cover in the Rechlin area. The result was a further two-week interruption to the flight test programme.

In view of this unsatisfactory situation, Prof Dr-Ing Heinkel requested via *Oberstleutnant* Petersen the immediate transfer of all flight-testing back to Rostock-Marienehe, because in his opinion this could prevent further delays. However, despite several petitions, everything remained as it was

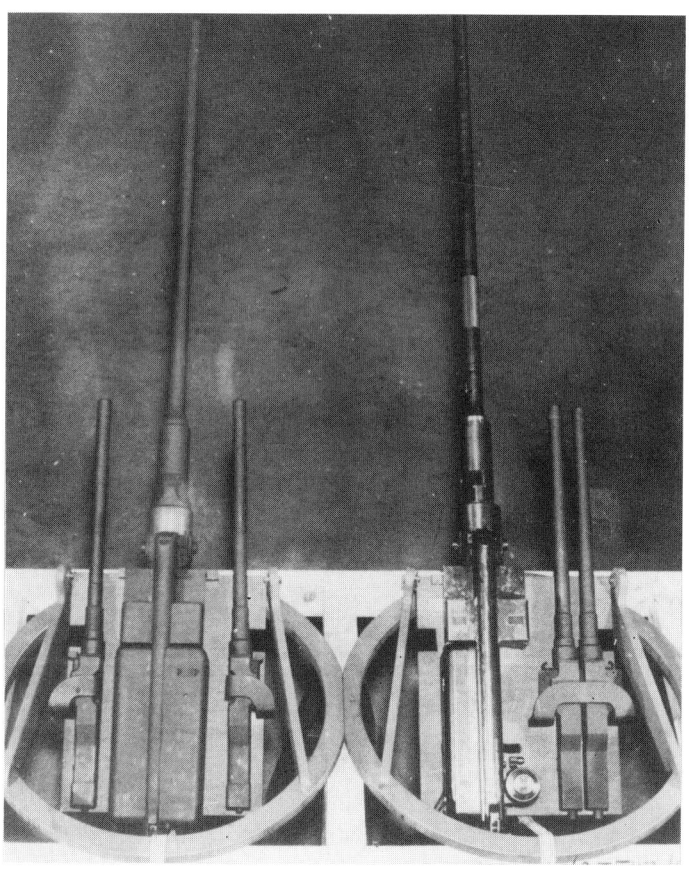

Left and opposite:
As on the Do 217 and Ju 288, various combinations of powerful mixed-gun defensive armament were tested on the He 177.

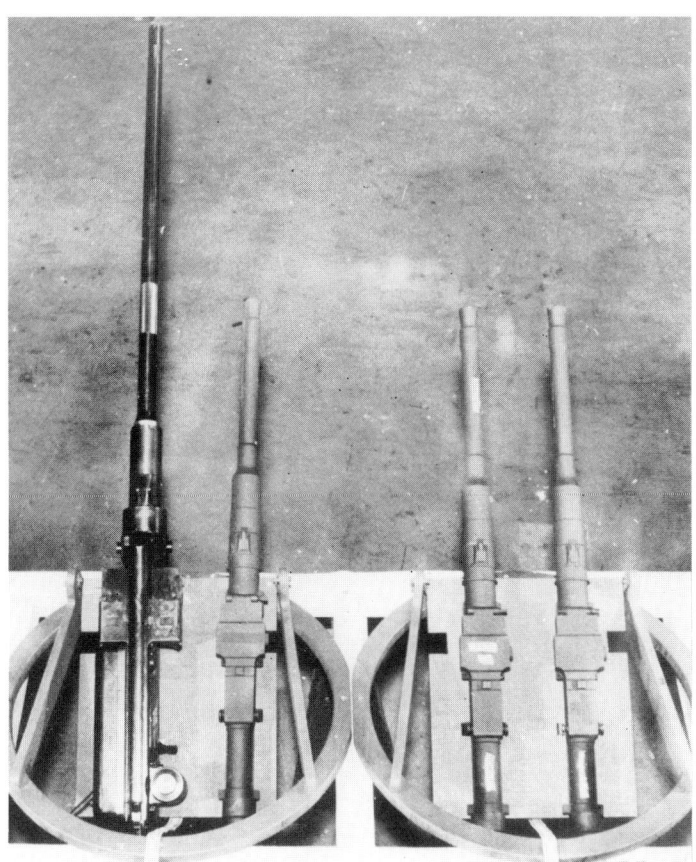

for the time being. Yet with fewer than six prototypes available, a new production-standard model could not be made operational for years – something only a few people within the RLM seemed to be worried about.

Defensive Armament

The expected defensive armament of the He 177 was laid down by the RLM in summer 1936, long before the configuration of future production series He 177s was even finalised. The aircraft was to defend itself against attacks from above, below and aft by means of two automatic weapons, and from frontal attack by another one or two automatic weapons. Initially, no thought was given to a tail defensive position. It was only after a requirement was formulated by the *Technische Amt* in 1937 that such a position was added to the design and incorporated in all subsequent He 177 proposals.

After the incorporation of a pressurised cabin into the He 177 design, the intended defensive armament was changed from manually-operated to remote-controlled 7.92 mm MG 81 and 13 mm MG 131 machine-guns, these being considered sufficient for defensive purposes. A functional mock-up of the He 177's cabin with manually-operated weapons was demonstrated in April 1939, at which time Dipl-Ing Francke expressed his wish for the design and construction of a considerably more spacious cabin. The result was that the initial He 177 prototypes remained temporarily unarmed or were fitted with considerably reduced defensive installations. Not until 1940 was a pattern aircraft, the He 177 V12 (GI+BL/Wk Nr 15151), built with a completely new cabin incorporating remote-controlled defensive armament.

On 26 April 1939 *Oberst* Wilk ordered that each defensive position should be equipped with one MG 131, while in his opinion a single 7.92 mm MG 81I or twin MG 81Z (*Zwilling*: Twin/Coupled)

A close-up view of the initially-fitted tail gun, usually a single-barrel MG 131A1.

Abb. 45: Heckstand
1 = Lafette
2 = Sichtteil
3 = Trittmuschel
4 = Einstiegklappe

A view of the tail-end position from below, showing the emergency fuel jettison arrangement.

Abb. 43: Heckstand von unten gesehen
Die abwerfbaren Klappen zum Schnellablaß sind abgenommen
1 = DBU-Stoffschläuche

could be considered as the most effective substitute weapons for each MG 131. But a problem arose with difficulties in securing deliveries of the intended weapons, particularly the MG 81s. To avoid this, in October 1939 Heinkel was ordered to fit MG 131s as substitute cabin armament on the He 177 prototypes. However, problems with the MG 131 had turned out to be more serious and complicated than at first expected, and so early He 177A-0 pre-production series aircraft being built by Heinkel and Arado had to be retrofitted with MG 81s, as the long-promised MG 131s would not be available for some time.

To complicate matters still further, the delivery dates for the nose and dorsal gun positions being manufactured by Siemens and Rheinmettal-Borsig respectively were constantly postponed, so that the fixed fly-off date of the fully-armed He 177 V6 in late July 1940 could not be kept.

As a result of the difficulties surrounding the supply of new MG 81s and MG 131s, the first He 177A-0 pre-production aircraft had to be fitted in part with the 7.92 mm MG 15 'emergency armament'. The defensive armament of the He 177 V6 and V8 (Cabin form 2) was similarly affected and was therefore revised by Heinkel to comprise the following:

- Nose   1 x MG 15 in Ikaria ball-mount
- Dorsal   1 x MG 131 (manually-operated)
- Ventral   1 x MG 15 in Ikaria ball-mount
- Tail   1 x MG 131 (manually-operated)

In contrast, the He 177A-02 and A-03 (Cabin form 3) featured a slightly improved defensive armament configuration:

- Nose   1 x MG 81I   for the observer (as substitute for a remote-controlled MG 131)
- Dorsal 1 x MG 81Z   in rotating gun-mount (as substitute for a remote-controlled MG 131)
- Ventral 1 x MG 81Z   in a roller gun-mount (as substitute for a remote-controlled MG 131)
- Tail   1 x MG 81I   manually-operated

The He 177 V6 was the first aircraft retroactively armed with one 20 mm MG FF cannon (in the lower nose position) and equipped with a partially-armoured tail position.

For the sake of uniformity, all He 177s were to receive the same FDL 131/2 (*Fernbedienbare Drehlafette*: Remote-Controlled Rotating Gun-Mount) barbettes as developed for the Ju 288; but as the delivery situation of completed machine-guns seemed extremely uncertain, it was intended to fit the 35 He 177A-0 pre-production and 100 He 177A-1 production series He 177s with alternative armament as a precaution. However, the requirement plan drawn up in May 1941 still envisaged the fitting of remote-controlled dorsal gun barbettes, beginning with the first He 177A-3 produced by Heinkel's Oranienburg works and the 21st He 177A-1 built by Arado. The first aircraft so armed were to be delivered in February (A-1) and March (A-3) 1942 respectively, although due to test flights with the He 177 V6 and V8 at Lüneburg beginning in July 1941, all trials with the improved armament had of necessity to be deferred.

On 13 August 1941 Prof Dr-Ing Heinkel wrote to General-Ing Lucht and reminded him of the still unsatisfactory defensive armament situation with regard to the He 177. Meanwhile, the He 177 had been flown by operationally experienced *Luftwaffe* crews who criticised the dorsal MG 81 installation on account of its limited effectiveness and poor field of fire due to the propeller blades. This proved the correctness of the earlier request at least to retrofit the planned remote-controlled MG 131s (which, according to the RLM, should have been delivered earlier on) in time for the bomber's imminent operational service.

After the demonstration of the He 177 V6, delivery of the remote-controlled dorsal and ventral gun installations was once again promised by the RLM and Dipl-Ing Francke (appointed by *Reichsmarschall* Göring as *Commissar* for weapons procurement), this time for August 1941. According to an RLM directive, the until-then postponed operational trials had to start as quickly as possible.

The He 177A-04 (GA+QM/Wk-Nr 00 0019) was selected as the test-bed for the remote-controlled defensive weapons and transferred to *E-Stelle* Tarnewitz on 28 October 1941. Ground- and air-firing trials were successfully completed by 12 November 1941, one of the recommendations being that, based on practical experience, the second (aft) dorsal position should without fail be equipped with a DL 131-I (*Drehlafette*: Rotating Gun-Mount).

34 Heinkel He 177-277-274

Along with the He 177 V8, the He 177A-04 was fitted with new, enlarged engine cowlings and radiators.

An illustration from the He 177A-1 manual, depicting the forward fuselage of the He 177A-03 (DL+AR/Wk-Nr 00 0018) with an MG FF cannon in the lower nose position.

After several petitions the RLM approved the substantially strengthened defensive armament configuration on all He 177s, with the following to be installed by 10 March 1942:

- Nose (upper)   1 x MG 131   manually-controlled
- Nose (lower)   1 x MG 81I   remote-controlled
- Dorsal (forward) 1 x MG 131   remote-controlled
- Dorsal (aft)   1 x DL 131   manually-controlled
- Ventral   1 x WL 131*   manually-controlled
- Tail   1 x MG 151   manually-controlled

* WL (*Walzenlafette*: Roller Gun-Mount)

THE HE 177 IS BORN   35

The aft dorsal gun turret of an He 177 A-3 in service with KG 40. This turret was fitted with an MG 131 I machine-gun as standard.

Close-up of the aft dorsal gun turret and its manually-operated single-barrel MG 131 I.

It was planned to deliver the first 20 He 177A-1 bombers with this defensive armament configuration by 20 March 1942. A little later it was decided to fit an additional 20 mm MG FF cannon in the nose; while one MG 81 in the ventral cupola and one remote-controlled MG 131 in the forward dorsal barbette represented a real improvement in defence capability. But there was a change of configuration: the aft dorsal position was now deleted; the ventral gondola was to have an MG 81Z instead of a single MG 131; and a manually-operated MG 131 was to be fitted in the tail position. However, the aft dorsal position with a DL 131 turret was reintroduced from the 61st He 177A-1. The RLM also proposed the use of the heavier 30 mm MK 101 cannon instead of the MG FF cannon in the flexible upper nose gun position.

After some time, technical trials of the He 177's defensive armament revealed that the original tail-end compartment was too uncomfortable for the gunner on longer flights.

The tail gunner of an He 177A-3 of KG 1 in the process of climbing into his turret. The weapon is an MG 151/20 fitted with a special night-sight.

A further improvement of the tail defence zone began in 1941 with the development of one 20 mm MG 151/20 in an FHL 151 (*Fernbedienbare Hecklafette*: Remotely-Controlled Rear Gun-Mount) fitted with a periscopic sight, and a corresponding mock-up was built in November of the same year. But this was only a temporary solution; at the same time, work was also underway on two much more powerful tail turrets: one armed with a pair of MG 151s, the other with a quartet of MG 131s. To begin with, only fully-functioning mock-ups of both these turrets were built.

The pattern aircraft fitted with an MG 131Z in its tail turret was the He 177A-1/V20 (VF+RD/Wk-Nr 15254), which arrived at *E-Stelle* Tarnewitz in June 1943. In addition to the new tail armament, this aircraft had one MG 131 in the upper nose gun position.

The first bench-firing trials with the HL 131Z tail turret, fitted on an He 177A-3, could only begin the following month, but numerous faults, particularly with empty cartridge disposal arrangements, prevented the expected quick success. A further delay was caused by cracks in the aircraft's powerplants, so that aerial gunnery trials did not start until late July 1943. Again there were problems with empty cartridge disposal,

Two views of the manually-operated tail turret equipped with a pair of MG 151/20 cannon.

38   HEINKEL HE 177-277-274

An MG 131Z in the tail gun turret of an He 177A-3 (6N+AS/Wk-Nr 332618) assigned to KG 100.

Maintenance work underway on the remote-controlled forward dorsal gun barbette armed with a twin-barrel MG 131Z. Note the sighting station further forward on the aircraft.

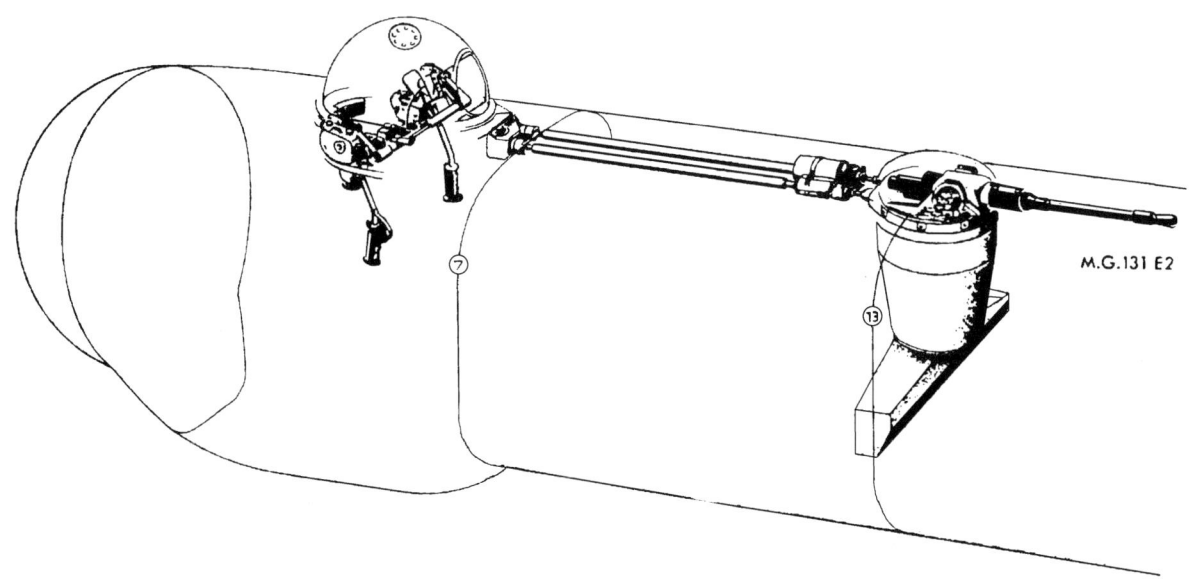

Schematic view of the remote-controlled forward dorsal gun barbette and attendant sighting station.

and the technicians did not expect any real remedy to be found until further tests had been conducted.

While these tests were underway, the second He 177A-3 pattern aircraft arrived at Tarnewitz. The first aerial gunnery trials using this aircraft took place in August 1943, but it was soon grounded due to a powerplant defect, and all further flight tests had to be put on hold. Beginning 22 August, a few acceleration flight tests to improve in-flight control were conducted, and all flight tests as such (except for altitude flights) were completed the following month, when the aircraft had to be grounded once again due to a control fault. Afterwards, it was fitted with an improved twin-gun tail turret.

By mid-September, both of the He 177A-3s at Tarnewitz had been fitted with the HL 131Z V1 and V2 turrets and were ready for further trials. The main difference between the two turrets lay in their armour protection, wherein the V2 was somewhat lighter in weight.

To begin with, the V1 had to undergo a series of bench-firing tests to finally resolve the problem of empty cartridge disposal. As the testing progressed, numerous defects were encountered with both turrets until October 1943, particularly during high-altitude firing trials and empty cartridge disposal, but certain alterations finally cleared these faults.

Until the end of 1943, practically all of the gunnery trials were undertaken using the second He 177A-3 to arrive at Tarnewitz. The aircraft went on to conduct tests lasting several weeks in 1944 in the Vienna area. In March 1944 the He 177 V38 (KM+TB/Wk-Nr 550002), the pattern aircraft with an MG 131 in both dorsal and ventral turrets and equipped with the FuG 200 *Hohentwiel* ASV radar, was also at Vienna. Practical trials using this aircraft began at Tarnewitz on 27 April 1944.

The use of the twin-barrel MG 131Z in the tail turret was favoured by Dipl-Ing Hertel, who envisaged four remote-controlled MG 131Zs in the turrets on his reworked He 177 design; only the tail armament remained manually-operated.

The second proposal featured a remote-controlled MG 131Z in the dorsal turret, two manually-operated MG 131Zs, and a quadruple-barrel MG 131V (*Vierling*: Quadruple) in the tail turret. This configuration corresponded to proposals regarding improved flexible armament worked out in October 1943 by *E-Staffel* 177 at *E-Stelle* Rechlin-Lärz. A study undertaken in February 1943 according to RLM instructions

The entry hatch into the quadruple-barrel HL 131V tail turret, pattern constructions of which were tested from July 1943 onwards.

Inside view of the rather cramped HL 131V tail turret, showing the oxygen supply installations for the gunner.

Two views of the HL 131V tail turret that did not enter series production. It was intended for most later-production He 177s up to and including the He 277.

envisaged the defensive armament of the entire He 177A-7 series consisting of quadruple-barrel HL 131Vs instead of twin-barrel MG 131Z turrets. Only a few weeks later it was decided to fit an HL 131V on the He 177A-6, with the HL 131Z turret to serve only as an interim solution.

From August 1943 onwards Prof Dr-Ing Heinkel planned to imbue the flight test programme with a greater sense of urgency, using the He 177 V32 (GP+WC/Wk-Nr 535353), V33 (GP+WD/Wk-Nr 535354) and V34 (GP+WN/Wk-Nr 535364).

A functional mock-up of the new, more powerful tail turret was to be ready at *E-Stelle* Rechlin by 1 April 1943. An experimental installation had been built and fitted on the He 177A-04 and ground-tested at *E-Stelle* Tarnewitz late in October 1941, but due to technical problems the manufacture of operational turrets was delayed until 1 April 1943. However, the He 177 V32, the first production series He 177A-3 pattern aircraft and armed with the HL 131V tail turret, did not arrive at Tarnewitz until early September 1943. Before that, there had been considerable delays at Rheinmetall-Borsig. As the reliability of the Daimler-Benz powerplants could not be guaranteed, a second aircraft, the He 177 V33, was flown to Tarnewitz on 2 September 1943. Meanwhile, the existing tail gun installation was removed in readiness for the next stage of the trials programme.

According to the evaluation programme, works tests with the quadruple-barrel 'HL 131 Vier V1' first prototype turret had to be completed, and simultaneously the 'HL 131 Vier V2' second prototype turret prepared for installation by the end of September 1943. In addition to certain alterations to the entry hatch cover, the functional tests had to be completed in just eight days. Aerial gunnery trials began early in October.

In November 1943, both the V32 and V33 were

grounded at the same time because their powerplants were losing too much oil. There were also problems with the gun installations, and the cartridge belt feed could not be made functional until mid-December.

Projects were also underway to arm some He 177A- and B-series aircraft with quadruple-barrel nose turrets. Although this gun installation cut speed by some 30 km/h (19 mph) and was responsible for a reduction of about 1,000 kg (2,205 lb) in offensive load, these remote-controlled nose turrets steadily grew in importance. The reason for this was the losses suffered on maritime-reconnaissance operations by the '*Kehl* and Atlantic' aircraft, which were becoming increasingly frequent victims of Allied long-range fighters.

Another experimental installation, a remote-controlled twin-barrel nose turret instead of an MG 151/20 in the lower nose gun position, was fitted on an He 177A-1 (GI+BP/Wk-Nr 15155). Practical tests began immediately, and were only briefly interrupted due to some damage to the aircraft's undercarriage. However, this installation was not considered for the new He 177A-3 and A-5 due to the required structural alterations.

Nevertheless, on request from the Chief of the Development Department of the RLM these nose turrets were to be installed on some He 177B-5 *Kehl*-equipped aircraft which would leave the production lines from December 1944. In the event, the cancellation of the planned He 177B-series meant that the entire twin-gun nose turret development effort turned out to have been in vain.

Trials with Drop-Loads

Project P 1041, planned by Siegfried Günter in 1936, was initially intended to carry a bomb-load of 500 kg (1,102 lb) and have an operational range of 2,500 km (1,553 miles). At that time the RLM still greatly valued 'large amounts' of 50 kg (110 lb) bombs carried in an internal bomb-bay. In fact the RLM did not see any special need for 500 kg (1,102 lb) bombs, although the carriage of such ordnance had already been intended by Heinkel. On 2 April 1938, however, the *Technische Amt* suddenly notified Heinkel that the minimum bomb-load had to be increased to at least 1,000 kg (2,205 lb). From then on, the drop-loads under

Standard frontal defensive armament aboard all He 177A-1 to A-5 aircraft: one MG 151 in the lower nose gun position.

A slight mishap with the He 177A-1/V29 (GI+BP/Wk-Nr 15155) during defensive armament trials at E-Stelle Rechlin.

consideration comprised bombs ranging from 10-500 kg (22-1,102 lb).

For the planned production series aircraft the Development Department of the RLM proposed the *Lofternrohr* (*Lotfe*) telescopic bomb-sight and the *Reflexvisier* (*Revi*) 12C reflex sight for 'inclined flight attacks'. The decision came two months later: the He 177A-0 bomb-aiming equipment would consist of the BZA 1 sight.

To help recognise difficulties during loading trials in good time, the manufacturers built a partial mock-up of the three-section bomb-bay. Late in 1939 the RLM stated that initially only SC 50, SC 250 and SC 500 bombs would be carried

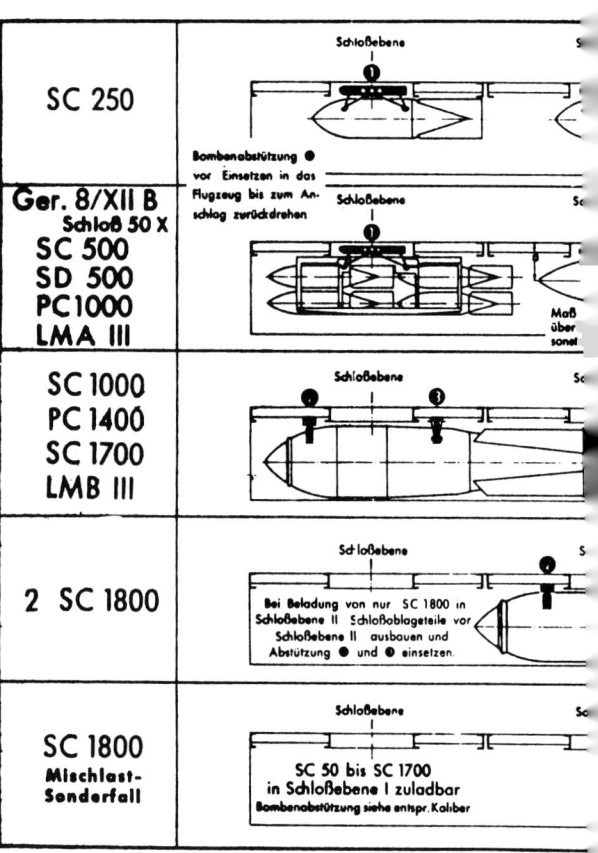

The Lofternrohr *(Lotfe) 7 telescopic bomb-sight, fitted to the He 177.*

in corresponding bomb-bays. Later, it was requested that consideration be given to the possibility of also carrying *Luftminen* (LM: Aerial Mines) and other large-calibre loads up to 66 cm (26 in) in diameter. Test aircraft used to evaluate the drop-load installations and a two-section bomb-bay were the He 177 V6 and V8.

The RLM representatives put great value on a quickly available, simplified drop-load release mechanism which was to be installed at Rostock-Marienehe only in the He 177 V4 and V5. In the meantime, the *Technische Amt* kept establishing ever more requirements which inevitably necessitated some redesign. For example, due to the demand that the aircraft be able to transport varied mixed loads, the electrical element of the release mechanism had to be considerably altered.

The installation of the selected bomb-aiming equipment in the He 177 was discussed during a conference at Rostock-Marienehe on 31 January 1940. On later pre-production and production series aircraft it was intended to fit the improved *Lotfe* 7B/7C and BZA 1 bomb-sights. For now, however, the Design Bureau saw no problems with the proposed *Lotfe* 7C AK bomb-sight, and the *Sturzkampfvisier* (*Stuvi*) 5 dive-bombing sight was similarly fitted inside the fuselage mock-up and fully accepted by the RLM representatives.

In March 1940, due to structural improvements to the PC 1000, SC 1000, PC 1700, SD 1700 and SC 1800 bombs the He 177's fuselage bomb-bay had to be modified again, resulting in a delay of several weeks.

During a two-day loading exercise in May 1940 the bomb-bay mock-up was first loaded with two SC 250s, then one SC 500 and one SD 500. The bomb-loads were 'dropped' by means of the emergency jettison equipment. The next load

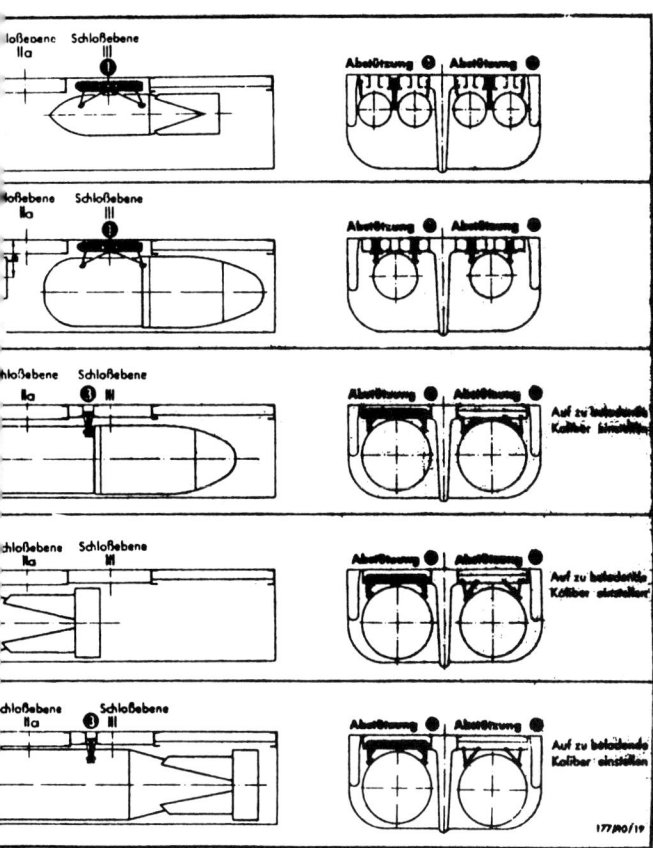

A chart of internal offensive load combinations approved for carriage by the He 177. The bombs range from the SC 50 to SC 1800.

consisted of the LMA III and LMB III aerial mines, one SD 1000 and a special 10-pack of wooden SC 50 mock-ups, all of which were attached in the bomb-bay and then lowered by means of a hoisting cable. In addition, trial attachments were made with one each of the PC 1000, SC 1000 II and SC 1800, as well as the *Lufttorpedo* (LT: Aerial Torpedo) 5.

The biggest problems were created by the Gerust 50/X frame equipped with the 10 individual SC 50s. The representatives from *E-Stelle* Tarnewitz were opposed to the existing configuration, describing it as impractical and demanding a structurally better design. This new 10-pack frame was to be tested at the end of September 1940, but as all later versions were also rejected by the RLM, SC 50s were not considered for operational use by the He 177.

The first prototype and pre-production aircraft intended for operational trials had the choice of the following drop-load combinations:

- 6 x SC 250
- 6 x SC 550/SD 550
- 6 x SD 1000
- 4 x SD 1400
- 4 x SC 1700
- 2 x SC 1800
- 6 x LMA III
- 4 x LMB III

Another ordnance-loading exercise with the He 177 V6 in November 1940 showed that the early prototypes could initially take only six instead of the 12 SC 250s planned for the pre-production series aircraft. On 24 February 1941 the He 177 V6 had to be transferred to *E-Stelle* Rechlin for trials with the bomb-release mechanism. Later, the aircraft was fitted with the revised flexible defensive armament.

During a tactical comparison between the Fw 200 Condor and the He 177, the latter gained higher marks. Despite the Fw 200's longer range, the He 177 had better all-round performance and could carry larger-calibre bombs on operations. Apart from that, by early 1941 an increasing number of Fw 200s were being lost to accurate

The He 177A-02 (DL+AQ/Wk-Nr 00 0018), which was still flying under KdE orders late in June 1943.

shipborne anti-aircraft fire. According to *Hauptmann* Diesings, the Fw 200 was simply too slow.

Meanwhile, trials were underway to perfect the drop-load release equipment on the He 177, and on 19 May 1941 the He 177A-02 arrived at *E-Stelle* Rechlin for further tests. After the compulsory acceptance control the aircraft made two works test flights on 24 May, followed by an emergency jettison test flight carrying six 1,000 kg (2,205 lb) bombs. Shortly thereafter, a powerplant defect provided a brief break in flying until 27 May.

On the fifth test flight the aircraft suffered a distorted bomb-bay door drive. Three days later, loaded with three SC 1000s and two SC 500s, only the latter two bombs left the bomb-bay when the emergency jettison mechanism was activated; the heavier bombs remained resolutely stuck despite repeated operation of the emergency release. By 4 June the aircraft had revealed other defects and was grounded at Rechlin, and despite the installation of several new press switches the emergency jettison equipment remained defective.

On 8 June the aircraft was flown to Rostock-Marienehe in an attempt to locate and rectify the faults. A long-range test flight with four SC 1000s the next day had to be aborted prematurely due to a defect in the hydraulic system. The following two test flights also ended prematurely, and a few days later the aircraft was once again parked on the Rostock-Marienehe works airfield, awaiting yet more repairs.

Subsequent He 177A-1 flight tests in late summer 1942 revealed deficient stability around the normal and lateral axes, resulting in extremely poor bombing accuracy when using the *Lotfe* 7D telescopic bomb-sight. The main reason for this was the drifting motion of the aircraft in flight due to its relatively short, round fuselage. This was only rectified on the He 177A-3 and A-5, which featured an elongated fuselage that gave much better stability and consequently considerably more bomb hits in the target circle.

Unloading of newly-delivered bombs. These three photographs provide a good view of the various transport crates and appliances used.

The official He 177 A-1 to A-4 Type Sheet.

By 1943, carriage of increasingly heavy drop-loads had priority in the He 177 development programme. Thus, the He 177A-7 Type Sheet of 13 August 1943 records one SC 2500 on an external fuselage bomb-rack and two SC 500s in the rear bomb-bay. The same bomb-load was also intended for the projected He 177A-6.

The actual operational use of the He 177's load-carrying capabilities began with the dropping of supply containers and the first bombing raids by aircraft assigned to I *Gruppe* (Wing) of *Fernkampfgeschwader* (FKG: Long-Range Bomber Group) 50 in the 'Fortress Stalingrad' area. This was followed by the first attacks on shipping targets on 23 August 1943 by aircraft of II/KG 40. During the following months, He 177s carried out numerous attacks on Allied shipping and landing troops. In summer 1944, He 177s were again operational on the Eastern Front, but these bombing raids had to be terminated after a few weeks due to heavy losses. The same applied to attacks on the Invasion Front, where the Allies had destroyed the German forces within a few weeks.

Chapter 2
The Start of Operational Service

A fine study of the He 177A-02, the second of 15 pre-production series examples built at Heinkel's Rostock-Marienehe works. The aircraft spent most of its life on trials work at E-Stelle Rechlin.

The Way to the He 177A-1

After the manufacture by Heinkel and Arado of the 35 He 177A-0 pre-production series aircraft, Arado went on to licence-build 130 examples of the first production series model, the He 177A-1, the first example of which (GI+BL/Wk-Nr 15151) was also designated the He 177 V12 for test purposes. Following factory flight tests at Brandenburg-Neuendorf, numerous A-1s subsequently underwent practical trials at Rostock-Marienehe, Rechlin-Lärz and, later, Vienna-Schwechat.

According to the RLM Production Summary of 1 August 1944, the only difference between the He 177A-0 and the A-1 was the six additional Type 2000 XIIIB bomb racks fitted to the latter. The A-1 was armed with one MG 81 in the upper nose and one MG FF in the lower nose, an MG 81Z with 1,000 rounds of ammunition in the ventral position, a remote-controlled MG 131 in the dorsal turret, and a manually-operated MG 131 in the tail turret. Radio equipment comprised the *Funkgerät* (FuG: Radio/Radar Set) 10, FuG 16, *Peilgerät* (PeilG: Direction-Finding Set) 2A and FuBL 1.

THE START OF OPERATIONAL SERVICE 51

Illustrations from the He 177 official handbook show the fin/rudder configurations up to and including the He 177 A-04 (left), and from the He 177 A-05 (GA+QN/Wk-Nr 00 0020) onwards (right).

A low-angle view of the He 177 A-04 (GA+QM/Wk-Nr 00 0019), showing to good effect the wing flap track.

Basically, the He 177A-1 was, and remained, only an interim model superseded in due course by the more efficient and effective A-3. Even so, over a period of several months there appeared more new problems, resulting in the A-1s on the production line requiring numerous modifications. These problems, some of which were only finally solved by summer 1942, included the following:

- January 1940: Fracture of main undercarriage cranking pin
  Wrong kind of rivets in engine shaft support bearer
- February 1940: Alteration of powerplant installation (following the crash of the He 177A-011)
- March 1940: Fowler flap drive and setting
- April 1940: Elevator and rudder mass balance (following the crash of the He 177 V3)
  Shearing of DB 606 gear bolts
  Strengthening of DB 606 engine coupling
- May 1940: Strengthening of DB 606 engine carriers
- June 1940: Misalignment of propeller setting
- July 1940: Enlargement of horizontal tailplane by 40 per cent
  Alteration of split flaps
- September 1942: Improvement of cold start equipment

As a result of these problems, only a few He 177A-1s could be used for type trials by the Heinkel works and the KdE until summer 1942. Those aircraft available were used as follows:

Operator	Flying hours	Tasks
Heinkel	430	Development/experiments
Heinkel	175	Acceptance/training
E-Stelle Rechlin	261	Trials (9 a/c)
E-Stelle Tarnewitz	44	Trials (7 a/c)
E-Stelle Travemünde	3	Aerial torpedo trials
E-Stelle Peenemünde	19	Hs 293 trials (1 a/c)
E-Staffel 177	372	Trials (8 a/c)

During these trials a total of nine He 177A-1s were lost: five at Heinkel, one at *E-Stelle* Rechlin, and three at *E-Staffel* 177.

The 26 prototypes accumulated a total of 1,291 flying hours, and evaluation of the data acquired during the flights led to 170 separate improvements being made to the He 177 airframe up to November 1941. In addition, no less than 1,395 structural alterations were recommended following the breaking load tests and strength evaluations at the Heinkel works. As a result of the constant alterations being made to production series aircraft, the frequent use of only conditionally cleared test aircraft and several bad accidents (which cost the lives of 17 aircrew), it was impossible to keep to any flight trials timetable.

In conjunction with the *Bauaufsicht Luft* (BAL: Construction Supervisor, Air), *Oberst* Petersen (KdE) submitted a comprehensive report to *Reichsmarschall* Göring on 13 August 1942, detailing the progress of the He 177A-1 tests to date. *Oberst* Petersen came to the conclusion that because of the 'underdeveloped and unreliable powerplants' no operations by He 177s worth mentioning could be expected before August 1943. Göring practically seethed with rage; the operational debut of the He 177A-1 once again seemed to have receded into the far distance:

'Why has this silly engine suddenly turned up, which is so idiotically welded together. They told me then, there would be two engines connected behind each other, and suddenly there appears this misbegotten monster of welded-together engines one cannot get at!'

Large-scale production of the He 177A-3 with DB 610 powerplants was supposed to start in October 1942, after the preliminary A-1 series with DB 606s, but the new engine had not even been test-run by then and the signs for the future did not look good: based on practical experience with the DB 606, it could be expected that due to the higher engine thermic temperature there would be difficulties with the cooling installation as well as oil circulation, and the technicians at Rechlin could already foresee a plethora of alterations.

In an attempt to speed up the trials programme as much as possible, the first eight He 177A-3s were to be delivered to Lärz in October 1943 for continuous rating tests of the DB 610. To ensure

Close-up view of the DB 610 coupled powerplant comprising two DB 605s.

that no He 177A-3s would be deployed for operations with underdeveloped powerplants, it was decreed that all A-3s delivered between October 1942 and January 1943 should instead be fitted with DB 606s.

Then, early in September 1942, there appeared new problems of a quite serious nature concerning the strength of the He 177's wings during inclined and diving flights. During tests, an He 177A-1 (BL+FU/Wk-Nr 15191) had revealed considerable wing deformation after just 20 flights. The deformation was particularly severe in the central area of the end spar, the outer wing sections along the wing leading-edge as well as near the airbrakes.

Prof Dr-Ing Heinkel's earnest warnings about persisting with the development of a diving capability for such a large aircraft had had no noticeable effect as yet. In fact, diving tests continued until the end of September 1942 with up to five aircraft, including the He 177 V5 and A-0/V15, the A-1 that had suffered the wing deformation, the A-1/V27 (V4+UC/Wk-Nr 15203) and another A-1 (VE+UP/Wk-Nr 15216). The latter three aircraft were later re-equipped (along with others) and redesignated as He 177A-3s.

A concerted effort by the Heinkel works led to an extensive (and expensive) wing strengthening programme that finally cured the wing deformation problem. Afterwards, seven consecutive diving flights showed clearly the success of this extensive structural revision.

In the meantime, new reports had given *Reichsmarschall* Göring another scare; for it seems that until summer 1942 he had believed the He 177 to be a four-engined aircraft, and one capable of diving attacks at that:

> 'I had told Udet from the start that I wanted this beast with four engines. This crate must have had four engines at some time! Nobody had told me anything about this hocus-pocus with welded-together engines.'

It must have been Prof Dr-Ing Heinkel's detailed letter to *Oberst* Vorwald, in charge of the *Generalluftzeugmeister Technische Amt* (GL/C: General of Air Production, Technical Office) at the RLM, that led to some serious contemplation regarding the He 177's powerplant problems and diving capability. Nevertheless, the RLM still would not relinquish its demand for 'full diving capability'.

A few days later, on 15 September 1942, Göring personally intervened in the matter and notified Heinkel via General-Ing Lucht:

> 'The Reichsmarschall has ordered that diving ability is not required of the He 177. He has rightly termed this demand as madness, and has forbidden it.'

From then on, it was a case of making up for lost time. The Heinkel works had to quickly do everything to make the He 177A-3 fit for operational service. This involved a total of 18 A-1 and A-3 pattern aircraft which had to be swiftly re-equipped and test-flown as follows:

- 8 to be rebuilt as *Zerstörer* aircraft
- 4 for inclined flight tests

- 1 for flights with all-up weight up to 33,000 kg (72,751 lb)
- 1 for radio equipment tests
- 1 for pressurised cockpit tests
- 1 for further armament tests
- 1 for control modification tests
- 1 for heating tests

However, as a result of ever more new demands from the RLM and the OKL (90 per cent) as well as the necessary improvements at the Heinkel works (10 per cent), a new problem had arisen: over the years, the He 177's all-up weight had increased to 34,000 kg (74,956 lb). For this reason the aircraft's internal fuel capacity had to be reduced, which, in turn, led to some reduction in the tactical range of the He 177A-1 to A-5 series. It was only after *Oberst* Petersen and Major Mons had become decisive backers of all matters relating to the He 177 that most of the technical problems were finally taken in hand.

Meanwhile, factory flight-testing of the He 177A-3 had commenced at Vienna-Schwechat (headquarters of Heinkel-Süd), whence the Heinkel workers had been transferred from Rostock-Marienehe after the RLM and the *Luftwaffengeneralstab* had finally agreed late in July 1942 to relocate the Heinkel design office and the manufacture of pattern aircraft in Austria.

Of special importance now were all the tests to ensure the full reliability of the aircraft. One of the test aircraft employed for this purpose was the He 177A-1/V25 (GI+BN/Wk-Nr 15153), which was used to evaluate a fuselage lengthened by 1.60 m (5 ft 3 in) and intended for the production series aircraft. This new, elongated fuselage offered a marked improvement in stability around the normal axes and was adopted for all subsequent He 177 models.

According to the 'Advisory Comments' of Dipl-Ing Hertel, a much more fundamental and far-reaching improvement of the He 177's operational reliability could be achieved by means of certain minor alterations to that most unreliable element of the He 177, the powerplant installation. He suggested fitting the following:

- More efficient fire extinguishers
- An interim piece of sheet metal above the spark plugs
- Elastic connections of pipe lines
- Larger and more powerful coolant pumps

Together with the improvements in stability brought about by the longer fuselage, these alterations to the powerplant installation did indeed increase the He 177's general safety in flight. In addition, the manufacturer paid particular attention to all possibilities that could potentially

He 177 Series Planning

Version:	A-1/R4	A-3/R4	A-3/R5	A-5
Powerplant:	2 x DB 606	2 x DB 610	2 x DB 610	2 x DB 610
Weight (loaded):	31,000 kg (68,343 lb)	31,000 kg (68,343 lb)	31,000 kg (68,343 lb)	34,000 kg (74,957 lb)
Weight increase due to:	Aft dorsal turret	*Kutonase**; MG 151/20 in nose & tail positions	*Kutonase**; final wing strengthening	*Kutonase**; full wing strengthening
Loss in fuel capacity:	-400 kg (882 lb)	-1,140 kg (2,513 lb)	-1,340 kg (2,954 lb)	—
Reduction in range:	-300 km (186 miles)	-750 km (466 miles)	-900 km (559 miles)	-700 km (435 miles)
Resultant range:	5,000 km (3,107 miles)	4,750 km (2,952 miles)	4,600 km (2,858 miles)	4,800 km (2,983 miles)
Delivery of first production example:	Summer 1942	Autumn 1942	Spring 1943	End 1943

* *Kutonase*: barrage balloon cable-cutter built into wing leading-edge

enhance and expand the bomber's overall performance and operational capabilities.

The first result of this was a long-range reconnaissance version with a range up to 6,000 km (3,728 miles), work on which began after a discussion with the RLM. It was intended to produce 40 such aircraft, with the first pattern aircraft to be built and fitted out at Oranienburg in collaboration with the *Versuchsstelle für Höhenflüge* (VfH: Experimental Establishment for High-Altitude Flights), the former *Kommando* Rowehl, by 1 September 1942. The aircraft (a modified He 177A-1) was to have its weight reduced by the removal of the MG FF cannon and its ammunition, and was to be fitted with the GM-1 nitrous oxide power-boost unit. The camera equipment was provided by the VfH and installed in the bomb-bay.

Although the performance estimates were promising, no series production of this long-range reconnaissance version was undertaken. Only a few modified A-1s were temporarily used by the VfH in support of the project, these including GI+BV/Wk-Nr 15161, VE+UJ/Wk-Nr 15210, VE+UK/Wk-Nr 15211 and VE+UL/Wk-Nr 15212. Of these four, the first crashed due to unknown causes on 2 September 1942, and shortly thereafter, the remaining trio were returned and reconfigured as bombers.

Operational Trials and E-Staffel 177

Examination of the mock-up of the He 177's bombing equipment was followed from 1940 onwards by practical 'operational trials', with blind- and night-flying trials for operations over Great Britain as the next priority. It was initially planned to prepare the He 177 V6-V8, A-02 and A-03 aircraft for this purpose by 1 March 1941. Following acceptance by the BAL, the aircraft would be immediately transferred to *E-Stellen* Rechlin and Tarnewitz for trials. But it was not to be.

Several months went by before the V6 could commence bombing trials in February 1941, and the installation of the MG FF cannon for use against ground targets alone took several weeks. That being so, soon nobody at KG 40 expected any operational trials to start before August 1941. In the event, instead of being used for bombing trials, the V6–V8 trio had to be transferred to

The He 177A-03, one of 35 pre-production aircraft built, commences its take-off run.

The He 177 V7 airframe undergoing static load-bearing tests.

Out in the open during trials at E-Stelle *Rechlin, the He 177 V7, partially covered by camouflage netting.*

Lüneburg for use in the familiarisation of *Luftwaffe* crews new to the He 177. A side-effect of this was a delay in the planned continuous rating tests of powerplants at Rechlin. To make matters worse, numerous unforeseen faults and technical defects required repeated treatment in the workshops, with the result that there was only modest progress in the He 177's trials programme as a whole.

In mid-August 1941 Prof Dr-Ing Heinkel reported rather optimistically to the RLM that achieving operational status with the He 177 now rested solely upon receiving the requested weapon installations, propellers and replacement powerplants. In reality, it was not until early October 1941 that the He 177 V8 could be delivered for tests at *E-Stelle* Rechlin, followed by the A-02 and A-03 by 20 October, by which time the whole trials programme was already well behind schedule.

By early November 1941, instead of the 18 He 177s which should have been completed according to the Delivery Plan, only three prototypes were available, and of those the V6 and V7 had been serving as training aircraft for KG 40 crews at Lechfeld since June. The first true 'operational' aircraft was to be the He 177A-07, expected to arrive at Rechlin in December 1941.

Meanwhile, plans to extend the operational range of the He 177 by somewhat unconventional means had reached an advanced stage. According to these plans, after climbing to a predetermined

The Start of Operational Service

The He 177A-08/V9 (GA+QQ/Wk-Nr 00 0023) during service with E-Staffel 177, *an operational trials unit established at* E-Stelle *Rechlin-Lärz on 1 February 1942.*

Initially equipped with a few He 177A-0s, E-Staffel 177 soon received He 177A-1s, including this example (VE+UO/Wk-Nr 15215).

Another view of VE+UO whilst operating from E-Stelle *Rechlin-Lärz.*

altitude the aircraft was to be refuelled in mid-air from a 'flying tanker', possibly a modified Ju 290. This would increase the He 177's range without bomb-load to at least 10,000 km (6,214 miles). A year later, this idea had reached fruition in the He 177A-2 project, a version featuring a pressurised cockpit and in-flight refuelling (IFR) equipment. According to project estimates, this high-altitude reconnaissance aircraft would have a range of 5,800 km (3,607 miles) on internal fuel alone, this being easily increased to 9,500 km (5,903 miles) by a single IFR connection.

Another version capable of IFR was the He 177A-1 *Zerstörer* (Destroyer/Heavy Fighter), intended for long-range maritime operations. As well as jettisonable fuel tanks carried in the bomb-bay, the Heinkel design bureau also proposed the use of two or four large external drop tanks. In the event, however, neither of these projected models became a reality.

E-Staffel 177, an operational trials unit under the command of Major Mons, was established at *E-Stelle* Rechlin-Lärz on 1 February 1942 and initially equipped with a few He 177A-0s, supplemented by some A-1s shortly thereafter. The He 177 had proved its flight endurance capabilities late in January 1942, when the A-07 piloted by Major Mons and manned by an operational

	Me 210A-1	Beaufighter 1F	Hurricane I	Spitfire IX
Powerplant:	2 x DB 610F	2 x Hercules IX	1 x Merlin III	1 x Merlin 61
Take-off Power:	1,395 hp each	1,500 hp each	1,030 hp	1,565 hp
Maximum Speed:	620 km/h	520 km/h	511 km/h	657 km/h
	(385 mph)	(323 mph)	(318 mph)	(408 mph)
Armament:	2 x 20 mm cannon;	4 x 20 mm cannon;	8 x .303 in MGs	2 x 20 mm cannon;
	2 x 7.92 mm MGs	6 x .303-in MGs		4 x .303-in MGs

THE START OF OPERATIONAL SERVICE   59

Luftwaffe crew had conducted the He 177's first 12-hour flight, covering a distance of 4,700 km (2,920 miles).

Comparison flight tests carried out against an Me 210A-1 and captured Bristol Beaufighter 1F and Hawker Hurricane I fighters had shown that the He 177 could hardly be overtaken and attacked. As far as speed was concerned, only the latest Supermarine Spitfire models had the necessary power to overtake the Heinkel bomber. Another air interception test was conducted on 14 May 1942, when *Hauptmann* von Holthey flew simulated attacks against an He 177 using an Fw 190 and Bf 109; afterwards Major Mons recommended that the MG 131 tail defensive armament should be improved as quickly as possible.

Since the beginning of the Second World War, vital supplies necessary for Great Britain had been delivered by a series of convoys traversing the mid-Atlantic. These were constantly attacked by U-boats, but apart from anti-submarine measures the Allies – with more than 500 maritime combat aircraft – had also succeeded in shooting down numerous Fw 200 Condors operated by KG 40. The faster and more powerful He 177, superior to the Condor, was now tasked to regain the lost ground.

At least one *Gruppe* of He 177s would be employed on anti-shipping tasks. In addition to strafing and the use of aerial torpedoes, it was planned to introduce the very promising LT 350 parachute-retarded torpedo. Also, the He 177 *Zerstörer* was expected to make its operational debut soon, and to use its 30 mm MK 101 cannon to effectively combat the British anti-submarine aircraft. So much for some of the purely tactical aspects.

In mid-1942 *E-Staffel* 177 possessed only two He 177 pre-production series aircraft. During the previous six months three of the eight available aircraft had crashed, and for that reason many of the outstanding tests could not be completed. Nevertheless, the RLM was convinced that the He 177 was 'an aircraft of high combat value, despite the still-existing faults'. This was rather surprising, because the performance check-flights carried out in the meantime had revealed the He 177 to be able to penetrate to a depth of only 2,100 km (1,305 miles), and to have a maximum speed of just 360 km/h (224 mph) when carrying a 2,000 kg (4,410 lb) bomb-load – all of which led *Oberstleutnant* Petersen to conclude that no He 177 maritime operations could be expected before March 1943.

There were several good reasons to support Petersen's claim. For one thing, due to the still-prevalent powerplant problems Major Mons had to caution against use of the He 177 over the wide expanses of the Atlantic Ocean. For another, checking the various alterations and changes requested to be carried out on production series aircraft up to that time alone would take several months of intensive work. Furthermore, according to Petersen, the re-equipment of a formation of *Gruppe* strength (approximately 24 aircraft in three *Staffeln*), including training to ensure the full flying and technical competence of the aircrews and ground personnel respectively, would take an additional six months. To make matters worse, *E-Staffel* 177 had not even managed to complete the glide and diving attack flight tests.

Apart from these obstacles, the KdE considered the development of a new cockpit as well as modified ailerons and elevators to be of special importance. There was also interest regarding an improved fuel transfer system, the moving forward of each powerplant by 20 cm (7.87 in) and the design of a new, strengthened wing that would withstand stresses of up to 4*g*. From the defensive armament point of view, the selection of a quadruple-barrel tail turret was at the top of the list.

However, as quite often nearly all the extant He 177s assigned to *E-Staffel* 177 were grounded at the same time, the necessary work was noticeably delayed. The situation at *E-Stelle* Rechlin in October 1942 was accurately described by Major Mons as follows: The characteristics of gliding attacks using the BZA sight were very good. It was possible to carry out attacks of up to 70° inclination for short periods, but due to insufficient wing strength attack flights were limited to 40° inclination. The He 177's behaviour in level attack was judged somewhat less favourably, although the elongated fuselage proposed for the production series aircraft would soon rectify that drawback.

From a flying point of view, low-level attacks were also well within the realms of possibility, although strongly-defended targets, especially those covered by light flak, would result in heavy losses because the He 177's powerplants, fuel tanks and crew areas were unprotected. Apart

A low-level flypast by VE+UO during its time with E-Staffel 177. Sadly this aircraft was lost as the result of a crash on 21 December 1942.

An He 177A-3 of E-Staffel 177 during engine endurance tests. Note the MG 81I on a flexible mount in the upper nose gun position.

from that, the He 177 as a whole represented a target that could hardly be missed.

According to Major Mons, the He 177's characteristics during single-engined flight and nocturnal take-offs were somewhat more acute than those of the Ju 88A-4. On the other hand, the bomber's behaviour during low-speed flight and immediately prior to touchdown with the Fowler flaps lowered was first-rate. For these reasons even average pilots would be able to hold the He 177 in a bank of up to 40 at a speed of approximately 190 km/h (118 mph). Care had to be taken not to undercut this speed, as had happened with two *E-Staffel* 177 aircraft. Both had suddenly stalled and crashed, with the loss of all on board.

Surprisingly, the DB 606 powerplant, which in the meantime had achieved a 'life expectancy' of up to 130 flying hours, was praised. The reason for this was the equal power-loading of both units, although their servicing and maintenance continued to be adversely criticised as so often before.

The first He 177A-3s to arrive at *E-Staffel* 177 were greatly praised, partly on account of their improved defensive armament consisting of two MG 151s, three MG 131s and one MG 81. Having

Two views of the remains of an He 177A-3 that crashed following failure of one of its coupled powerplants.

said that, the A-3 also required considerably greater maintenance effort compared to other *Luftwaffe* bomber types. It took a lot more work to keep the A-3 serviceable than, for example, the Fw 200 Condor; in addition to which the He 177 also required highly-qualified technical support personnel. Thus, for example, the A-3 had to undergo an accurate and exacting 25-hour servicing after each long-range flight. The powerplants required special attention, as did the susceptible electro-hydraulic system, the electric wiring system, and the remote-controlled gun installations which had to be checked at short intervals. The same degree of attention was later also applied to the He 177A-5 with its troublesome *Kehl* radio-guided bomb control equipment.

So much for the opinion of Major Mons, which primarily concerned itself with the technical and tactical aspects of the He 177 trials.

Apart from *E-Staffel* 177, work also continued at an undiminished pace at *E-Stelle* Rechlin. The He 177A-3/V24 (ND+SS/Wk-Nr 135024), fitted with improved DB 610 powerplants, was first thoroughly tested in June 1943. Although this work resulted in a broken propeller, and shortly afterwards, the failure of one of the new DB 610s, the aircraft was not seriously damaged.

A new major test effort began with the arrival at *E-Stelle* Rechlin of the first two He 177A-5s, these being rebuilt A-3 airframes. The first pattern aircraft, the He 177 V22 (VF+QD/Wk-Nr 332104), landed at Rechlin in July 1943. Follow-up final check-flights by Rechlin test pilots showed that the aircraft was in order, apart from the poor return movement of the ailerons. However, flights on one engine could only be safely carried out at an all-up weight of no more than 22,500 kg (49,603 lb). It was also established that when flying at combat power the engine oil temperature rose quite critically due to the old-type cooler installations and the high summer temperatures.

When the He 177 V22 had to undergo another powerplant replacement, it was temporarily transferred to Tarnewitz. Afterwards, the aircraft was used for a diverse range of performance measurement flights to establish the range table for the He 177A-5. Delays caused by several oil leaks led on 24 September 1943 to a second rebuilt He 177A-3 (NN+QS/Wk-Nr 535552) having to be collected from the Heinkel works to enable the measuring flights to continue. These flights were followed in October 1943 by flight tests on one engine with fixed *Elektrische Trägervorrichtung für Cylinderbomben* (ETC: Electrically-Operated Carriers for Cylindrical Bombs) as well as complete exhaust flame damper attachments. These tests established a maximum permissible continuous flying speed of 255 km/h (158 mph) on one coupled powerplant at sea level.

Several test flights with external loads of up to 1,800 kg (3,968 lb) in November 1943 also failed to provide any negative results. These flights were in fact the preliminary experiments for the subsequent attachment of guided 'special weapons'. To conclude their work, the Rechlin *Abteilung* E7 (Section E7) carried out loading and dropping experiments with SC 2500 bombs in December 1943.

A series of overload take-offs at all-up weights up to 33,000 kg (72,751 lb) were also performed with an He 177A-1 (GI+BX/Wk-Nr 15163) in conjunction with *E-Staffel* 177. Another A-1 at Rechlin was used for the final inspection of the FuG 200 *Hohentwiel* and FuG 216 radar installations. Prior to that, both aircraft had been brought up to A-3 equipment standard.

E-Staffel 177 was disbanded at the end of 1943. Some of its personnel were transferred to *E-Stelle* Rechlin for other tasks, while some technicians were posted by the *Luftwaffenpersonnelamt* (Air Force Personnel Directorate) to KG 40.

He 177 Production

On 17 October 1938 a conference was held at the RLM which determined Germany's entire aircraft production plan for the period from July 1940 to February 1942. According to the plan, production of He 177s at Heinkel's Rostock-Marienehe works would be discontinued after completion of the 31st aircraft in December 1940, three months after licence-production had commenced at two other plants in September 1940. Production of the He 177A-1 was set at 25 aircraft per month from April 1941 onwards.

During the first licence-production conference on 15 August 1939, the RLM announced that the following firms had been earmarked to undertake He 177 series production:

- Ernst Heinkel Flugzeugwerke (EHF) at Rostock-Marienehe
- Heinkel-Werke GmbH at Oranienburg (HWO)
- Weser Flugzeugbau GmbH (WFG) at Bremen

At that time, production of the He 177 still did not merit any special 'urgency' within the aircraft production programme; the Ju 88 alone held first place. Soon there was also a change in the companies chosen to undertake the licence-production work, with Arado-Werke at Brandenburg (ARB) being taken on as a substitute for WFG. However, when the details of Delivery Programme No 18 were subsequently revealed, the He 177 was suddenly no longer listed under HWO and ARB. Until then, the Arado works alone had employed 500 men to build jigs to facilitate He 177 production, and these were now 80 per cent complete. In contrast, only about 100

THE START OF OPERATIONAL SERVICE   63

Series production of He 177 airframes together with coupled powerplant units at Heinkel's Rostock-Marienehe works.

He 177A-3 forward fuselage sections during construction.

In the background, an He 177A-1 from the Arado production line has its compass calibrated on the compass point at E-Stelle Rechlin.

personnel had been engaged in He 177 production at HWO.

The reason for the omission of HWO and ARB was because licence-production was to restart only after the completion of Delivery Programme No 18, on 30 August 1941. The interruption was caused by the limited stocks of cast steel. However, due to the material turnover at EHF – the highest among Germany's aircraft manufacturers – there were no significant material reserves, so that any interruption of licence-production at HWO and ARB could not be made good with EHF's own resources.

The continuous re-evaluation of the He 177's importance, a process mirrored by the RLM's classification of the aircraft, had a very disturbing effect on the overall development programme and created delays in the supply of both materials and equipment. Indeed, it was only in July 1940 that the RLM allocated the highest special priority rating to the He 177 and put it on the same level as the Ju 88. However, only a short while later, on 20 August 1940, came Hitler's instructions governing the execution of the special-priority production programme, which listed the He 111 but not the He 177.

Exactly one month later, the new long-range bomber was allocated the very low 'Priority 1A' rating. The suspension of series production for three months after two crashes probably had something to do with this decision. What was more certain was that the allocation of low production priority status came at just the right time for Dipl-Ing Lucht in the RLM, who was biased in favour of the Junkers concern and so now did not need to vigorously promote the He 177.

The whole situation changed again on 9 November 1940, when the State Defence Council allocated 'Special Priority' to He 177 production for a few weeks' duration. According to this revised status, the first production series aircraft were now to start leaving the ARB assembly lines as quickly as possible, with five He 177A-0 pre-production series aircraft to be followed by the first 60 A-1s by March 1942. Due to the short-sighted earlier interruption of licence-production, production at HWO could not start before October 1941, and so output totalled just 35 aircraft by March 1942 – just over a year after the He 177 had finally been allocated the highest 'Special Priority' rating on 7 February 1941.

By then a lot of time had been irretrievably lost, and despite every effort the He 177A-01 could not make its first flight before April 1941. Four months later, after some delay, the first He 177A-1

The Start of Operational Service 65

An early-production He 177A-1 photographed while serving with KG 40 at Châteaudun in 1943.

left the ARB production line. The main reason for this hesitant production was the still underdeveloped DB 606 powerplant, too few of which had been delivered in the first place. Apart from that, the He 177 airframe required some major and numerous minor improvements, all of which had to be incorporated as 'series alterations' on the production line. When added to the required strengthening of the airframe itself, the result was a lot of lost production time.

After delivery of the 35 He 177A-0s had been completed, production switched to the A-1, with the work being carried out exclusively by Arado at Brandenburg-Neuendorf and Warnemünde. In all, 130 A-1s were built by Arado. *Four sub-models were developed (the A-1/R1, R2, R3 and R4), all of which were heavy bombers differing primarily in the configuration of their defensive armament.* Twelve aircraft were further developed as A-1/U2 *Zerstörers*, easily identifiable thanks to the large cannon mounted in the lower nose position.

By June 1942, due to the slow start-up of production and some material and equipment shortages, there was a production shortfall of no less than 144 He 177s. Although it was known in Berlin that the RLM's production quotas could not be met, on 26 June 1942 the monthly production rate for the He 177 was again fixed at 120 aircraft. At this stage, *Generalfeldmarschall* Milch proposed to halt production of the He 177 at EHF and concentrate the final assembly at the HWO and ARB plants. However, by the end of June 1942 only 82 of the required 130 A-1s could be completed by the latter two production facilities.

Despite the already existing delivery shortfalls, the Special Production Committee F3 insisted in July 1942 on an immediate start to production of the He 177A-3, this to total no less than 2,228 aircraft. Of these, only the first 15 would be fitted with DB 606 engines; the rest would be powered by DB 610s, even though testing of the latter powerplant still had not been completed. But the intended early start to production of the A-3 was not to be, and the same happened to the planners three months later, leaving their hoped-for output of 120 A-3s per month as mere wishful thinking. In fact *only 335 A-3s were built, production comprising six sub-models (A-3/R1, R2 and R6 heavy bombers; A-3/R3 and R4 glide-bomb carriers; A-3/R7 torpedo-bomber).* A *Zerstörer* sub-model, the A-3/R5, remained as a project study only. In addition to the new-build aircraft, some A-1s were later upgraded to A-3 standard. Development and production of the projected A-2 and A-4 high-altitude bombers were not proceeded with.

By way of a remedy, production of the He 111 was cut back on 28 August 1942, to provide more workers for the He 177 production programme. However, the problem was not so much a matter of too few airframes as the ever-increasing

This He 177A-3 crashed during a flight to Brandis.

shortfall of DB 606 powerplants; to the extent that the entire programme was once again called into question. To make matters worse, tests had shown the He 177's wings to be below the strength required by the RLM for inclined flight, making time-intensive strengthening measures necessary.

Despite every effort, the He 177 was not as straightforward to build as, for example, the He 111, and only 107 examples were completed by 30 September 1942 – well below the planned number. At times, there were even interruptions in final test flights when new faults in the powerplants revealed themselves. As a result, the first He 177A-3s could not be fitted with DB 610 powerplants before October 1942, later than planned.

It was not until the end of November 1942 that the He 177 production figures finally matched the planned output. According to *Generalfeldmarschall* Milch, the Allies' mass production of aircraft had to be countered by German aircraft of superior quality, and that applied especially to the He 177 and Ju 288, which would later carry the main burden of the *Luftwaffe*'s offensive operations. *Reichsmarschall* Göring agreed with this opinion; to rely solely on the fast fighter-bombers for close support did not seem to be a practical solution. Yet despite this, the new Planning Study 1014 still envisaged a partial reduction in bomber production in favour of single-engined fighters and fighter-bombers, although there were no planned significant changes in the He 177 production programme.

Deliveries of He 177A-3s with DB 610 powerplants were to commence in January 1943, starting with 32 aircraft per month and rising to 100 aircraft per month by November 1943, and continuing at the latter rate until the end of 1945. The reality proved to be somewhat different.

Due to the protracted delivery of powerplants the entire He 177 production programme was slowed down yet again, so that only 38 aircraft could be expected in February 1943. The shortfall of necessary powerplants already amounted to 200 units in January 1943, and increased still further the following month. As airframe production was now running to plan, ever more He 177A-3s had to be parked engineless and uncompleted on factory airfields. To complicate matters still further, the DB 610 had revealed some serious faults which now put large-scale series production of the He 177A-3 in doubt.

It was at this stage that Dipl-Ing Heinrich Hertel wrote to *Oberst* Vorwald at the GL/C and expounded the current situation regarding production of the He 177 thus: The repeatedly demanded monthly production of 110 He 177A-3s beginning in April 1944 would only be achieved by taking immediate urgent measures. Beginning July 1943, aircraft with forward-set powerplants would start leaving the assembly

Final assembly of He 177s underway at the Eger Flugzeugwerke plant near Pilsen in occupied Czechoslovakia.

lines in noteworthy numbers. Hertel expected an end to the parking of engineless new aircraft from September 1943 onwards – assuming that Daimler-Benz could finally deliver a sufficient number of reliable powerplants.

All this was fine on paper, but it did not work out in practice. Three plants were engaged in retrofitting He 177A-3s with strengthened wings and moving the powerplants forward by 20 cm (7.87 in): the repair depots at Eger and Erfurt, and the Lufthansa workshops at Berlin-Tempelhof. Despite this arrangement, these measures did not have the desired effect as quickly as had been expected; on average, less than 30 He 177A-3s could be delivered per month between March and May 1943. In the meantime, 70 He 177A-3s had been parked on the Hagenow airfield because their engines were needed for fighter production. There had also been several casualties during the factory flight tests, while improper servicing did the rest.

Early in 1943 up to 26 faulty DB 610s per week were being exchanged by each He 177-equipped *Staffel* (average strength seven or eight aircraft). In fact, the whole situation could not have been more unsatisfactory, and the He 177's operational debut in the West had to be postponed month after month.

He 177 Production

Prototypes	40[1]	
A-0	35[2]	Produced at EHF and ARB
A-1	130[3]	Produced at ARB only; one pattern aircraft built by Heinkel-Ost
A-2	—	No production
A-3	335*	At least 146 produced at ARB and 24 at HWO
A-4	—	None produced
A-5	565**	Produced at ARB and HWO
A-6	15•	Rebuilds of A-3/A-5s
A-7	11••	Rebuilds of A-5s
A-8	—	None produced
A-10	—	None produced
B-5	4	Rebuilds of He 177A-series parts
B-7	—	None produced

[1] Apart from eight aircraft, all others rebuilds of A-0/A-1/A-3/A-5s by Heinkel AG
[2] According to other sources, only 27 A-0s built
[3] According to other sources, 149 A-1s built
* According to other sources, 171 A-3s built
** According to other sources, 789 A-5s built (at least enough sub-assemblies for that number were produced)
• According to other sources, six A-6s built
•• According to other sources, five A-7s built

Inspection of the twin-wheel main undercarriage units at Eger before the attachment of the outer wing sections and the four-bladed propellers.

On 27 June 1943, by special arrangement, Hitler received representatives of the seven leading German aircraft manufacturers at the Berghof, his mountain retreat on Obersalzberg in Bavaria. Mistrustful of the information supplied to him by Göring and his vassals, Hitler had invited the aircraft producers at short notice, so as to prevent Göring and Milch from intervening. His request was straightforward:

> 'I would like to put myself in the picture regarding the aero-technical situation; one that is not falsified by the coloured discourses of gentlemen from the Luftwaffe.'

Shortly afterwards, Hitler brought the conversation round to the He 177:

> 'This machine has been promised again and again, for years. When is the He 177 coming?'

Prof Dr-Ing Heinkel referred to the initially-demanded diving attack capability – but that had been expressly renounced by Göring on 15 September 1942, a good nine months ago. After some discussion about the *débâcle* of the coupled powerplants, Hitler suddenly and surprisingly completely changed the topic of conversation.

Back again at Rostock-Marienehe, consideration had to be given to how to fit the parked He 177s with DB 610s. The Fighter Programme needed 900 additional DB 605s, which meant the withdrawal from the production schedules of 225 He 177s between August 1943 and July 1944. As an interim measure, Prof Dr-Ing Heinkel intended temporarily to fill these gaps with reserve and rebuilt/repaired powerplants.

A meeting between Göring, Milch and the Daimler-Benz director Nallinger on 7 September 1943 became a rather heated affair when the subject of the He 177's powerplants was raised. After some fierce criticism the discussion casually turned to the DB 613 coupled powerplant (two DB 603s), two examples of which had already been built. Five more DB 613s were to be built by April 1944 for use in the test programme. However, none of those present at the meeting dared to make a prediction regarding the possible replacement of the DB 610 by this new coupled powerplant. In the end, an He 177 pattern aircraft was nevertheless fitted with two DB 613s and test-flown.

Only 20 He 177A-3s were completed in August 1943. By that time, the shortfall of powerplants amounted to 800 units; a figure large enough to effectively kill off any chance of reducing to any

In February 1944, because they could not be fitted with the necessary operational equipment and transferred to operational Luftwaffe *units in France, many newly-built He 177s were parked on the airfield at Fassberg in northern Germany.*

noticeable degree the number of engineless He 177A-3s, even with the supply of reserve/reconditioned powerplants. As a result, numerous engineless He 177A-3s remained parked where they stood. Meanwhile, Production Programmes 225 and 226 envisaged the following:

- Each batch of 100 He 177s would now consist of 70 'normal' bombers and 30 *Kehl*-equipped aircraft.
- From October 1944 onwards, Arado was to produce 20 He 177s completed to Equipment Condition 'C' (long-range bomber) per month until further notice.
- Simultaneously, HWO and ARB were to deliver 60 He 177A-5s per month. Both plants had to produce a total of 860 A-5s between them, of which ARB's share would amount to 452 aircraft.
- Production of the A-5 would terminate in favour of the He 177B-5 in March 1945. The first B-5s were to leave HWO in November 1944 and ARB in December 1944.
- From June 1945 onwards, it was planned to deliver 50 He 177s a month to the *Luftwaffe*. Production of the B-5, powered by four separate DB 603s, was to continue until March 1946, and total 2,450 aircraft.

Apparently hardly any of the participants drawing up these production programmes questioned where the high-grade raw materials required for these large aircraft were to come from.

In the event, all of these plans remained on paper only. On 25 May 1944 a counsel regarding the He 177's new defensive armament took place at Obersalzberg involving *Reichsmarschall* Göring, *Generalfeldmarschall* Milch, *Generalleutnant* Galland (GdJ), General Korten (Chief of the *Luftwaffengeneralstab*) and General Vorwald (GL/C), as well as *Oberst* Knemeyer (GL C/E) and Prof Dipl-Ing Tank (Technical Director of Focke-Wulf). Before that, *Reichsminister* Prof Albert Speer's *Jägerstab* (Fighter Staff), established within the *Reichskriegministerium* (State Ministry of War) on 1 March 1944, had expounded the jet-powered Ar 234, Ju 287 and Me 262 as the new standard aircraft of the *Luftwaffe*. The He 177 was still in the production programme, but with a planned monthly output of just 20 aircraft. Eight other aircraft types were to be deleted from production altogether: the Bf 108, Bf 110, He 219, Ju 52/3m, Ju 188, Ju 288, Ju 352 and Ju 390.

An He 177A-5 shortly before take-off from the Eger repair depot on its way to Brandis.

Just one month later, on 25 June 1944, came the turn of the He 177B-5. In his instruction accompanying its cancellation, Hitler added:

'The herewith associated renunciation of the strategic Luftwaffe for many years to come has to be accepted.'

On 1 July 1944, on the advice of the *Jägerstab*, *Reichsmarschall* Göring once again put to work his red pencil and deleted no less than 20 aircraft types from the production programme. One of those deleted was the He 177.

On the evening of 1 July – less than two months after Göring had pledged an increase in He 177A-5 production to 200 aircraft per month – the leading figures of the Heinkel works learned that the entire He 177 production programme had been dropped; a temporary closing down of bomber production for the benefit of the Bf 109 and Me 262. The already-established sub-assembly groups would be used in July and August for the final assembly of 40 He 177A-5s per month, with the final 33 aircraft to be assembled at HWO in September 1944. As replacement for the He 177 work, the HWO plant received a production order for the Dornier Do 335 *Pfeil* (Arrow). The first of these high-speed single-seat fighter-bombers, distinguished by their tandem fore-and-aft engine configuration, were to leave the HWO assembly line in February 1945, with output to rise to 200 aircraft per month from October 1945.

A corresponding phasing-out of He 177 production would take place at Arado. The freed capacity would be allocated to fighter production as well as the Ar 234 *Blitz* (Lightning) single-seat twin-jet reconnaissance-bomber.

While the news was still being digested, on 8 July *Reichsminister* Prof Albert Speer, Minister of Armaments and War Production, officially informed all concerned that Hitler had given his full agreement for the phasing-out of production of the He 177, and that an additional 1,000 fighters could be produced instead.

And so it came to pass that in August 1944 the final He 177A-5s of the already-manufactured assembly groups were completed and delivered to the *Luftwaffe*.

He 177 Production
(Excerpt from General Quarters 6th Department Production Statistics)

1942
January	5
February	1
March	7
April	13
May	17
June	10
July	17
August	16
September	16
October	18
November	19
December	31
TOTAL	**170**

1943
January	33
February	27
March	14
April	36
May	38
June	42
July	28
August	20
September	36
October	47
November	49
December	45
TOTAL	**415**

1944
January	59
February	64
March	86
April	71
May	77
June	86
July	95
August	5
September	22
October	—
TOTAL	**565**

| **GRAND TOTAL** | **1,150** |

Chef TLR	FLUGZEUG-BAUREIHEN-BLATT He 177				Chef TLR Fl.Nr.8555/44 gKdos(E-2) 550 Ausfertigungen 1.8.44

7.B1.

Baureihe	Triebwerk	Bewaffnung – Beladung	Abwurf Anlage	Kraftstoff	FT-Gerät	Sonstiges	Bemerkung
He 177 A-0 (K)	DB 606 A/B	MG 81 1000 Sch / MG 131/2000 Sch / MG 131 1500 Sch / MG FF 300 Sch / MG 81Z 1000 Sch / Bombenraum **Rüstzustand A (Nahbomber):** Beh.1 u.6 je 1140 l ⟶ Grundausrüstung; Beh.2 u.3 je 620 l Beh.7 u.8 je 1120 l } dazu austauschbar Beh.4 u.5 je 1520 l entspr. Rüst. ABC Ges.Inh. = 8800 l / Bomben = 7 t **Rüstzustand B (Mittelbomber):** Grundausrüstung wie Rz.A dazu wird Beh.4 von 1520 l ausgetauscht gegen einen Beh. von 3450 l. Ges.Inh. = 10 730 l / Bomben = 4 t **Rüstzustand C (Fernbomber)** Grundausrüstung wie Rz.A dazu wird Beh. 4 u.5 ausgetauscht gegen einen Beh. von 3450 l. Ges.Inh. = 12 660 l / Bomben = 1 t	6 Gerüst 8 Schloß 50 B-1 oder 12 Schloß 500/XII oder 6 Schloß 1000 XI b	Beh.1 = 1140 l Beh.2 = 620 l Beh.3 = 620 l Beh.4 = 1520 l Beh.5 = 1520 l bis 3520 l Beh.4 u.5 austauschb. entspr. Rüstzustand A/B/C. Beh.6 = 1140 l Beh.7 = 1120 l Beh.8 = 1120 l	FuG 10 FuG 16 Peil G 5a FuBl 1	Führersitz B-1 Stand Heckstand Bodenwanne Schlauchboot-wanne gepanzert	
He 177 A-1 (K)	DB 606 A/B	" "	zusätzlich 6 Schloß 2000 XIIIB	"	"	"	
He 177 A-3 (K)	DB 610 A/B	MG 131Z/2000 Sch / MG 131/1000 Sch / MG 131 1500 Sch / MG 81Z / MG 131/20 / MG 131/1000 Sch / MG 131/800	zusätzlich 2 ETC 2000 XII D mit Schloß 3000 IIIA bzw. Schloß 2000 XIIB in Schloßbla-kette	"	FuG 10P FuG 17Z FuBl 2P FuG 25a FuG 101 FuG 102 FuG 200 FuG 216 (FuG 203b)	zusätzlich B-2 Stand ge-panzert. Zusätzlich Rüstsatz E (Kehl Ausrü-stung)	
He 177 A-5 (K)	DB 610 A/B	MG 81/... MG 131Z/2000 Sch / MG 131 1000 Sch / MG 131/... MG 131/1000 Sch / MG 81/800 Sch	wie A-3	wie A-3	FuG 10P FuG 17Z FuBl 2P FuG 25a FuG 101 FuG 102 FuG 200 FuG 216 (FuG 203b)	wie A-3 zusätzlich Rüstsatz D (LT)	

A diagram showing the main equipment and armament differences between the He 177 A-1, A-3 and A-5 models as of 1 August 1944.

The He 177 as a Training Aircraft

At an early stage in the history of the He 177, in addition to the A-1, A-3 and A-5 versions, it was intended to produce a completely separate series of training aircraft with dual controls, but this idea had to be abandoned due to the somewhat limited production of the bomber.

Later, when initially only the He 177 V6, V7 and A-05 were available for crew training, the lack of regular production series training aircraft became all the more obvious. A small series of rebuilt aircraft based on an He 177A-0 (DR+IJ/Wk-Nr 32001) did little to relieve what was an increasingly strained situation.

The first *Gruppen* to start retraining on the He 177 were I and II/KG 40 at Fassberg and Lechfeld respectively in summer 1942. Prior to that, in August 1941, the He 177 V7 had flown for a few weeks with IV/KG 40 in France. Though short, the V7's time with IV *Gruppe* had generated a lasting interest in the He 177.

In 1943 the retraining of KG 4 'General Wever' and KG 100 'Wiking' crews at Lechfeld began on

These He 177 A-1s belonged to the complement of FFS (C) 5 at Neubrandenburg.

After being disposed of by an operational unit, this He 177A-3 (VD+XS/Wk-Nr 332143) served for a while with FFS (B) 16 at Burg bei Magdeburg. On 21 December 1943, the aircraft was damaged in an emergency landing.

This He 177A-3 too served with an operational unit before being handed over for aircrew training purposes. The emblem on the nose is that of FFS (B) 16.

a *Staffel*-by-*Staffel* basis. The only aircraft available for this purpose were mostly He 177A-0s and A-1s as well as a few A-3s, but there were no special training aircraft fitted with dual controls.

Early that same year, however, a training aircraft based on the A-1 was built and inspected at Ludwigslust, where it had been modified. It was intended to produce 83 such aircraft with reduced defensive armament and modified controls by the end of May 1943; but Heinkel's proposal to simultaneously modify 40 A-3 bombers found no response from *Generalfeldmarschall* Milch. As things stood, the production capacity was not sufficient to ensure an uninterrupted supply of aircraft to front-line units.

The modification process was slow: by 22 April 1943 only two He 177s had been rebuilt. To speed up things, *Feldwerftabteilung zur besondern Verwendung* (*Feldwerft-Abt zbV*: Field Repair Detachment for Special Duties) 1 of I/FKG 50 had

Clearly recognisable as a training aircraft, thanks to the large '32' on the rudder, this He 177A-3 was used by FFS (B) 16 at Burg bei Magdeburg.

been withdrawn from operational service and deployed for He 177A-series aircraft modification and re-equipment purposes. This move soon brought results, and by 28 June 1943 more than 20 newly-built training aircraft were on hand at Ludwigslust. Conversion work on another 11 trainers and a few prototypes was simultaneously underway at the Eger repair depot, also commissioned to undertake this work in summer 1943. Meanwhile, I/FKG 50 had more than 15 so-called 'newly-built training aircraft' on strength and in July a few more were modified at the Efurt repair depot.

After final flight tests some of these training aircraft were delivered to flying training schools in Germany and France. Thus, in summer 1943 *Flugzeugführerschule* B16 (FFS (B) 16: Advanced Flight Training School No 16) at Burg bei Magdeburg operated the first He 177A-1s in the training role. Most of the aircraft used for conversion training by FFS (B) 16 were previously assigned to KG 40. A few He 177A-1s and A-3s were also flown for a short time by FFS (B) 15 at Bourges in France, where two of the A-1s were badly damaged during an air raid on 10 April 1944. In May, FFS (B) 31 at Brandis lost three of its A-1s to enemy action in one raid.

In mid-1944 all He 177s still operated by FFS (B) 15 were handed over to FFS (B) 16 as part of an attempt to get a grip on the myriad servicing and maintenance problems still affecting the aircraft. Many crashes (the majority of which ended without any loss of life) had been caused by differential loading of powerplants, overstressing of airframes, and in some cases, insufficient instruction in the peculiarities of the He 177.

Despite its progress to date, FFS (B) 16 was closed down in July 1944. All 17 of its He 177s were grounded the following September due to a lack of aviation fuel. The aircraft park at Burg bei Magdeburg included six A-1s, nine A-3s, and two A-5s. Of these, the first to be scrapped were the Λ-1s; their powerplants were removed by Daimler-Benz technicians, assisted by the airfield maintenance personnel, and prepared for preservation and storage.

In the meantime, IV/KG 1 'Hindenburg', IV/KG 4 and IV/KG 100 had been combined to form *Ergänzungs-Kampfgruppe* (Erg-KGr: Reserve/Replacement Training Bomber Wing) 177 at Neuburg-Donau. As with FFS (B) 16 at Burg bei Magdeburg, all flying operations had ceased at Neuburg-Donau, so Erg-KGr 177 merely marked time before it too was disbanded.

Then, early in November 1944 came an order from the OKL to transform Erg-KGr 177 into IV

(Erg)/KG 40, consisting of a *Stab* and 10. and 12. *Staffeln*. This new training unit was given the task of preparing KG 40 crews for operations with the Me 262. However, due to limited deliveries of this novel twin-turbojet fighter/fighter-bomber, as well as Allied air raids on German jet aircraft production facilities, IV/KG 40 never received any Me 262A-1s.

A student crew at Burg bei Magdeburg preparing for a training flight aboard one of FFS (B) 16's He 177A-3s (VD+XS/Wk-Nr 332143).

Initial Operations of I/FKG 50

I/FKG 50 was formed at Brandenburg-Briest in summer 1942. The first He 177A-1s for 1. and 2./FKG 50 arrived in June, and Major Schede was appointed *Gruppenkommandeur* (Commanding Officer). Two months later, the RLM assigned Major Schede the practical trials of the new He 177 *Zerstörer*. Before that, Major Schede had been given the task of co-ordinating the strengthening of the wings of all delivered operational He 177s, while fitting of the new horizontal tailplanes took place not at Brandenburg-Briest but the Heinkel works at Rostock-Marienehe. At that time, Major Schede, *Gruppenkommandeur* of this the first He 177 operational unit, was promised by the *Luftwaffenführrungsstab* that a sufficient number of training aircraft would be made available for his crews.

However, as the He 177A-1 still needed various additional modifications, Major Schede had to report to the KdE, *Oberstleutnant* Petersen, that the 'He 177 is not suitable for operational service in its present form'. Worse still, 12 long-range flights under operational conditions had established a depth of penetration of only some 1,800 km (1,118 miles), and the average speed of 340 km/h (211 mph) was some 50 km/h (31 mph) less than that achieved by He 177A-0s flown by *E-Staffel* 177. Several crashes due to unexplained causes, such as that of an He 177A-1 (VD+UG/Wk-Nr 15232) on 14 November 1942, simply added to the bomber's negative image.

Nevertheless, the establishment of 3./FKG 50 was completed by mid-November 1942, while 4.(Erg)/FKG 50 was formed simply 'on orders from above' by the end of that same month. The necessary maintenance personnel were provided by *Feldwerft-Abt zbV* 1, which was subordinated to I/FKG 50 with effect from 20 November and likewise transferred to Brandenburg-Briest. In the meantime, Major Schede had taken care to ensure that the airfield intended for trials operations, Zaporozhye in southern Ukraine, was suitably prepared for the acceptance of 12 He 177A-1s.

By early December 1942, I/FKG 50 had a total of 33 He 177s, of which only 10 aircraft, i.e. just one-third of the effective unit strength, were serviceable. With these aircraft the crews had managed to make 347 take-offs and complete 290 flying hours to date. Bombing training (100 flying hours) and conversion training (89 flying hours) had accounted for most of these flights. Yet despite this commendable effort, of the 27 available aircrews only nine were fully conversant with the He 177; most of the crews had been unable to complete their bombing training due to the prevailing bad weather conditions. New difficulties were also expected due to the slow delivery of DB 606 powerplants. To ensure engine supply the RLM ordered the removal of powerplants from 50 He 177A-1s at Lüneburg, these to be stored as a readily-available reserve for I/FKG 50 aircraft.

Major Schede expected his unit to be operationally ready by January 1943, and the first trial operations began soon afterwards. Fighting in the Stalingrad area demanded massive air support for the German 6th Army from December 1942 onwards, and for that reason I/FKG 50 was subordinated to *Luftflotte* 4 in the VIII *Fliegerkorps* (Air Corps) region.

On 1 January 1943 *Feldwerft-Abt zbV* 1 left Brandenburg-Briest by train and arrived at the Zaporozhye-South airfield, eight kilometres (five miles) from Zaporozhye, 10 days later. Snowdrifts up to three metres (10 ft) high had paralysed all movement within the selected operational base and only the most important roads could be kept open by scores of shovelling Soviet prisoners-of-war. At night the temperature dropped to –35°C (–63°F), and was only a little

A line-up of ex-operational and very weathered He 177s relegated to training tasks with FFS (B) 16.

The funeral service for a I/FKG 50 crew killed when their He 177 crashed.

higher by day, all of which made the unloading of material a tortuous experience. For the want of an unloading ramp it took days before the train was emptied and the material transported to the base. Of the 15 hangars on the airfield, *Feldwerft-Abt zbV* 1 could use only three – sufficient to accommodate just five He 177A-1s.

By early January 1943 the detachment had established itself and received the first powerplants removed from He 177A-1s and dismantled at Lüneburg. Unfortunately, due to careless packing some parts and spares had not only been damaged but delivered without any consideration for the really important material.

The delivery of additional operational aircraft from Brandenburg-Briest also did not proceed without certain problems. On one He 177A-1 the connecting rod in the No 4 engine punched its way through the casing and the powerplant was soon ablaze. Fortunately the coolant tank above the powerplant burned through, allowing the coolant liquid to extinguish the flames. Another He 177A-3 with its share of misfortune (E8+FK/Wk-Nr 15252), piloted by *Oberleutnant* Lawatscheck, landed at Zaporozhye on 13 January 1943 after a malfunction of the fuel feed system and problems with de-icing of the horizontal stabiliser.

Despite every effort, conditions at the operational base barely improved. Due to the very deep snow and high snowdrifts it took half a day to move and arrange the He 177s, 'despite the assistance of a tractor as well as people'.

By 15 January 1943 a total of 26 He 177s had been transferred from Brandenburg-Briest, several of which had to return home prematurely; one aircraft crashed while attempting an emergency landing near Lemberg. By then the actual operational strength of I/FKG 50 was just 17 aircraft. Within a few days nine DB 606s had been completely wrecked and had to be dismantled. In some of the engine fires smoke had penetrated into the cockpit and hindered the pilot's reading of the flying instruments.

The Start of Operational Service

This I/FKG 50 He 177A-1 (BL+FJ/Wk-Nr 15180) crashed on 29 January 1943, with fatal consequences for all of the aircrew.

For the time being, the aircraft of I/FKG 50 were to be used exclusively for dropping 'supply bombs' to the troops of the German 6th Army, then surrounded in Stalingrad. But disaster struck on 16 January 1943. On that day five He 177A-3s led by Major Schede took off on a supply mission. Two aircraft returned early due to technical problems and malfunctioning engines, but all trace was lost of Major Schede's aircraft (E8+FH/Wk-Nr 15233). In the event, only one of the five aircraft managed to reach Stalingrad.

With Major Schede lost, *Hauptmann* Schlosser took over command of I/FKG 50, and the supply operations continued. The next day, five of the 28 He 177s at Zaporozhye were readied for another supply-dropping missions, but only two of the aircraft reached their destination. One aircraft was lost when one of its powerplants caught fire about 100 km (62 miles) west of Stalino, forcing the crew to bale out shortly afterwards.

On 19 January, eight He 177s took off on another supply-dropping mission, but due to

The He 177A-1 of Hauptmann Lange on the snow-covered Zaporozhye-South airfield in southern Ukraine.

poor visibility none of the crews found the agreed drop zone and all had to turn back. While landing at Kirovograd, *Oberleutnant* Lawatscheck's He 177A-3 (VD+UP/Wk-Nr 15241) clipped the hangar roof and ripped off the tailwheel. Despite this, the aircraft took off from the snow-covered runway the following day to return to Zaporozhye, only for it to crash on landing, four of the crew being killed as a result.

Only a few of I/FKG 50's He 177s now remained in more or less airworthy condition. Of these, two were loaded up with food supply containers on 22 January 1943 and flown to Stalingrad. On the way, both were intercepted by Soviet fighters and badly damaged. One of the aircraft (VD+UO/Wk-Nr 15240) received 18 hits

Due to the inhospitable winter weather conditions at Zaporozhye-South, serviceability of I/FKG 50's He 177A-1s and A-3s was frequently quite low.

The Start of Operational Service

He 177s of I/FKG 50 during supply operations to relieve the German 6th Army in Stalingrad during January 1943.

This He 177A-1 also belonged to I/FKG 50, and was used to drop supplies to German troops besieged in Stalingrad.

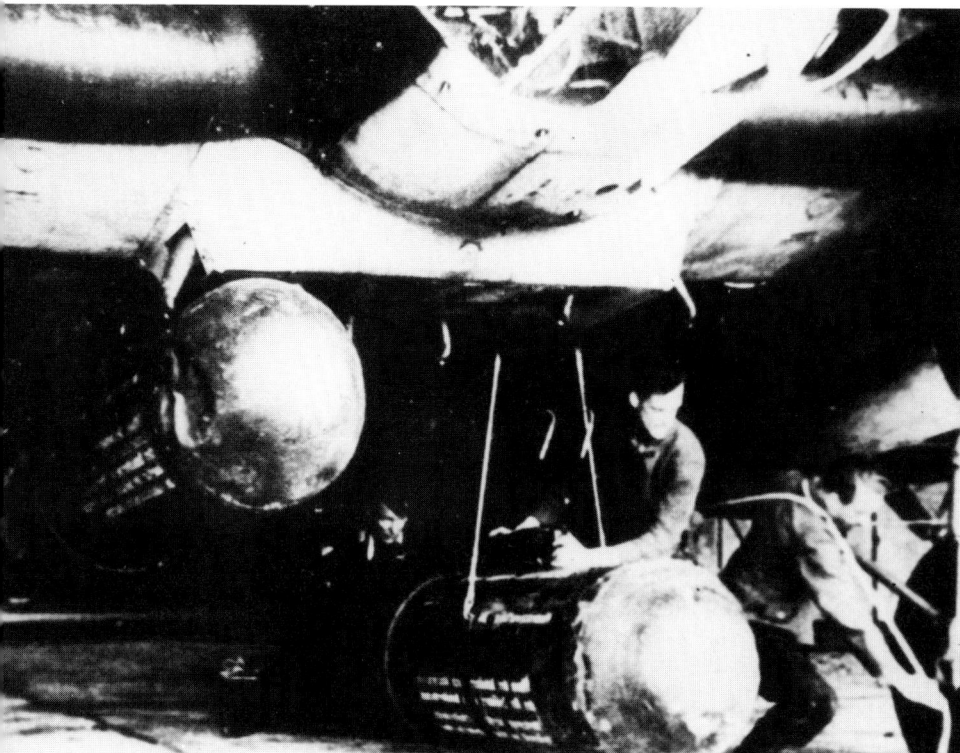

The type of supply containers air-dropped to the trapped German forces through the winter of 1942-43, seen here being attached beneath an He 111.

in the Fowler flaps, armoured ventral cupola, tail turret, both propellers and the rudder, yet still managed to reach the airfield at Kalinovka. On the second aircraft two of the four engines were shot out, making repairs impossible. Both crews reported the first sightings of Soviet-flown Spitfires, some of which had not yet been repainted and still bore RAF roundels.

On 23 January 1943 I/FKG 50 lost the He 177 piloted by *Oberleutnant* Spohr, the aircraft going missing after dropping its supply containers over Stalingrad. Two other aircraft turned back before reaching the drop zone, while another aircraft collected a bullet hole in its horizontal stabiliser. The next day, I/FKG 50 was directly subordinated to *Luftflottenkommando* (Air Fleet Command) 4; from then on bombing raids to bring relief to the German troops in 'Fortress Stalingrad' took priority over supply drops.

On 25 January, after a short pause due to unsuitable weather, five He 177s of I/FKG 50 and 12 Ju 88As of KG 51 'Edelweiss' carried out a successful raid in the Pitomnik area, dropping 128 SD 50 bombs for the first time. A day later the He 177s bombed Soviet forces in the Gumrak area, while a single aircraft carried out an armed reconnaissance along the Stalingrad-Dubovskoye railway line. Afterwards, numerous He 111H-6s dropped their 'supply bombs' over Stalingrad.

On the nights of 25-26 and 26-27 January 1943 a total of 171 *Luftwaffe* aircraft dropped supplies over the northern and southern pockets held by the surrounded German troops. A day later there followed an air raid by three He 177s on the western edge of 'Fortress Stalingrad', but a further four bombers had to turn back early due to engine problems. A similar fate befell I/FKG 50 on 29 January, when only half of the He 177s dispatched on a bombing mission managed to reach the target area.

On 30 January 1943 the *Luftwaffenführungsstab* received an urgent teleprinter message from *Generalfeldmarschall* Milch:

'Considering the six bad accidents involving the He 177s of I/FKG 50 within 14 days, I request the withdrawal of this formation from operations and transfer back to homeland. Signed, Milch, GFM'*

Despite Milch's request, the air supply operations in the Stalingrad area continued unabated. The participating crews reported 'large conflagrations and heavy artillery fire in the southern pocket'. On 31 January 1943 *Armeeoberkommando* 6 (AOK 6: High Command of the 6th Army) transmitted its final signal: 'We are destroying'. The battle for Stalingrad was over. Altogether the *Luftwaffe* lost 269 Ju 52/3ms, 169 He 111s, 46 Ju 86s, nine Fw 200s, five He 177s and one Ju 290 in the Stalingrad area.

I/FKG 50 was duly deployed back to Brandenburg-Briest. A comprehensive report analysing the *Gruppe*'s technical and tactical experiences during initial operations with the He 177 was then prepared and submitted by *Hauptmann* Schlosser, stating among other things that '... the aircraft had proved itself exceptionally well from the tactical point of view.'*

According to *Hauptmann* Schlosser, the high speed and the possibility of gaining higher speed quickly by pushing the He 177's nose down deserved special recommendation. The defensive armament, particularly after the fitting of the aft dorsal turret, gave the He 177 a good chance to repulse even strong enemy fighter attacks. For example, on 18 January 1943 an He 177 was attacked consistently by up to 14 single-engined Soviet fighters over a period of 40 minutes, during which time one of the He 177's gunners managed to shoot down one of the attackers and damage two others so badly that they had to break off their attacks. It was also evident that bombing accuracy had markedly improved since the end of 1942. Thanks to the *Lotfe* 7D telescopic bomb-sight, under normal conditions it was possible to score hits in a 35 m (115 ft) circle from an altitude of 3,000 m (9,840 ft).

Overall, *Hauptmann* Schlosser approved the He 177 for daylight operations, but also pointed out that certain technical problems, particularly those to do with the powerplants, had to be solved at any cost. The He 177 was not yet suitable for nocturnal operations due to the lack of engine exhaust flame dampers. On the other hand, special praise was reserved for the strength and ruggedness of the airframe and the undercarriage.

* Statement by *E-Staffel* 177, 5 April 1943

On 4 February 1943 *Feldwerft-Abt zbV* 1 received orders to deploy back to Germany. Despite this, six days later General-Ing Weidlinger, commander of *Luftflotte* 4, suddenly ordered that the unit now had to service and maintain Ju 88s at Kalinovka, even though the unit's personnel were only familiar with the He 177. The fact that two other field repair detachments in the same area specialised on Junkers aircraft made no difference. General-Ing Weidlinger stuck to his opinion that this particular unit was temporarily no longer needed to look after He 177s, as due to the bomber's shortcomings no operations worth mentioning could be expected before 1944!

February 1943 also saw the start of another chapter in the operational life of the He 177, when aircrews and technical personnel began a specialised instruction course at Brandenburg-Briest, followed soon afterwards by the arrival of the first three He 177 *Zerstörer* aircraft intended for planned 'train-busting' operations. In the event, nothing came of it: the operations on the Eastern Front, planned for summer 1943, finally had to be abandoned due to the unreliable Daimler-Benz powerplants.

On 28 June 1943 I/FKG 50 disposed of more than 20 He 177A-1s, some of which had been brought up to A-3 standard. In addition, the still-available *Zerstörer* aircraft armed with the 30 mm MK 101 cannon were used for training purposes. The unit remained busy on training duties for several more months, until on 25 October 1943, after detailing some crews to the *Flugzeug-führerschulen* C (FFS (C): Flying Training Schools, Multi-Engined Aircraft), it became the nucleus of II/KG 40.

Operations of KG 1 'Hindenburg'

Formed near the town of Holberg in 1939, KG 1 'Hindenburg' at first consisted of just a *Stab* and I *Gruppe*, both having been formed from the *Stab* and I/KG 152 'Hindenburg'. II/KG 1 came into being through the renaming of I/*Lehrgeschwader* (LG: Operational Development Wing) 3, previously IV/KG 152 'Hindenburg', while III/KG 1 was established from elements of I and II/KG 1 at Burg bei Magdeburg on 19 September 1939. KG 1 was expanded in summer 1940 through the addition of IV Gruppe as a training and replacement formation.

At the beginning of the Second World War in September 1939, *Stab* and I/KG 1 He 111Hs operated over Poland under the command of 1. *Fliegerdivision* (Air Division). In May 1940 KG 1's He 111Hs participated in *Fall Gelbe* (Case Yellow), the Western campaign in the Low Countries and France; and then in the costly *Adlerangriff* (Eagle Attack) raids on Great Britain starting in August 1940, the He 111Hs of *Stab*, I and II *Gruppen* being joined by the Ju 88A-1s of III *Gruppe*, all operating as part of *Luftflotte* 2. Daylight missions against Great Britain were augmented by the night-time '*Blitz*' bombing raids from September 1940 to May 1941, by which time KG 1 had been redeployed further east prior to the start of Operation *Barbarossa*: Germany's all-out advance against the Soviet Union.

In June 1941 I, II (also now Ju 88A-equipped) and III/KG 1 began operations against the Soviet Union as part of *Luftflotte* 1. The offensive operations on the northern sector of the Eastern Front continued until early January 1942, but soon afterwards the *Stab* and II and III *Gruppen* were recalled to Germany, leaving only I/KG 1 in the East for the time being. The *Stab* and III *Gruppe* returned in due course, but their efforts in support of attempts to cut through Soviet Army forces encircling the German 6th Army in Stalingrad were ultimately in vain, the southern group surrendering on 31 January 1943 and the northern group capitulating on 2 February 1943.

In spring 1943, KG 1 was re-equipped with new Ju 88As and crews at Neuausen and returned to service under *Luftflotte* 1 in May of that year. Two months later the *Stab* and I and II *Gruppen* were subordinated to II *Fliegerkorps* of *Luftflotte* 2 and redeployed to Italy; while during the summer months small elements of III/KG 1 began operations under the command of *Luftflotte* 6. Having participated in the battle of Stalingrad, there now followed attacks on Allied warships and merchant vessels in the Mediterranean Sea and Atlantic Ocean. Not before 1944 would KG 1, by then equipped with He 177s, again be operational on the Eastern Front.

On 18 November 1943 I/KG 1 began its transfer to Brandis for re-equipment with the He 177, and the entire *Gruppe* had completed its move by early December. Personnel were given rest leave until 6 January 1944 when the retraining was scheduled to start.

The first nine He 177A-1s (including some brought up to A-3 standard) landed at Burg bei Magdeburg in early March 1944, but the celebration of their arrival was short-lived. The aircraft turned out to be completely 'worn-out' examples

A KG 1 crew captained by Huber which took part in the costly operations on the Eastern Front.

THE START OF OPERATIONAL SERVICE   85

An He 177A-3 parked at Burg bei Magdeburg, with its dorsal machine-guns removed for cleaning.

formerly operated by KG 100 as well as some from II/KG 1. A further 20 He 177A-1s were allocated later that month, although 12 of these could not be ferried to Burg bei Magdeburg because of various defects.

Things began to look up soon afterwards, and by 31 March 1944 I/KG 1 had taken delivery of the first 24 He 177A-3 operational aircraft; six additional A-3s followed in April. After handing over nine A-1s and five A-3s to IV *Gruppe*, however, I/KG 1's establishment was down to just 22 He 177s by the end of April 1944.

To give an impression of the flying capabilities of the He 177 to the new, mostly very young aircrews, a group of Heinkel works pilots demonstrated the aircraft to KG 1. Demonstration flights by experts was one thing; the training of inexperienced aircrews on this heavy and complicated aircraft was quite another.

I/KG 1 suffered its first total loss during retraining at Brandis on 15 March 1944, when *Feldwebel* Krempler crashed his He 177A-1 (DH+CW/Wk-Nr 15197). This was followed by the forced landing by *Leutnant* Schneider of an He 177A-3 (V4+GH/Wk-Nr 332471) on 20 March. In April, I/KG 1 lost eight A-1s and A-3s: of these two crashed in flames early that month (A-3 V4+HL/Wk-Nr 332475 on 1 April, A-1 V4+UC/Wk-Nr 15203 on 5 April); two A-3s were lost due to pilot error (Wk-Nr 332478 on 17 April, V4+BK/Wk-Nr 332511 on 29 April); and another two A-3s made belly landings (Wk-Nr 332477 on 8 April, Wk-Nr 332410 on 13 April). In the latter two incidents, the reasons were the loss of powerplant and instruments respectively. Two remain unaccounted for.

By then I/KG 1 had a total of 46 aircrews on strength, but they were only partly retrained on the He 177. There was also a lack of well-trained technical ground personnel, such that a declaration of operational readiness could not be expected before 1 July 1944. The actual number of available aircraft was not a problem. Thus, on 21 December 1943 KG 1 had only two serviceable He 177A-3s; by the end of February 1944 the number had increased to 27 aircraft, and one month later, to 59 A-3s, 23 of them with II/KG 1. However, only 13 of the latter were serviceable. According to a report submitted to the *Luftwaffenführungsstab* on 14 April 1944, II/KG 1 at Burg had 16 He 177A-3s left, of which only three were cleared for operations.

Numerous major and minor faults regularly reduced the number of serviceable aircraft available. Among other things, there was a lack of equipment for the jacking-up of aircraft; cranes to facilitate powerplant exchange; vehicles to transport the DB 606 units – even loading machines for the filling of 13 mm MG 131 machine-gun and 20 mm MG 151 cannon cartridges into their disintegrating metal link belts. Until then, the armourers had loaded these belts by hammering in the cartridges one by one. Then there was the question of replacement powerplants. As only four units had been delivered for about 60 He 177s, the number of operational aircraft fell quite noticeably.

An He 177A- (VF+RP/Wk- 15266) of 6./KG 1 during retraining on the type in Germany in 1943.

Maintenance work underway on a 5./KG 1 He 177 A-3.

Finally, the operational readiness of the aircrews. While the combat-experienced crews of II/KG 1 flew the He 177 without difficulty, the newcomers had not yet mastered the Ju 88, never mind the much more complicated He 177. As a result, seven aircraft were lost during the re-equipment and retraining phase in May 1944. Two A-3s were damaged by enemy action while on a training flight on 21 May, one of which (KP+PN/Wk-Nr 332364) crashed after sustaining extremely heavy gunfire damage. Prior to that, on 10 May, one A-3 (Wk-Nr 332389) became unser-

The Start of Operational Service

The last offensive operations by He 177s on the Eastern Front were conducted by the A-3s of I and II/KG 1 operating from bases in East Prussia during summer 1944.

viceable due to a broken main undercarriage and another was lost due to pilot error; and on 15 May another A-3 (Wk-Nr 332539) became unserviceable due to tyre damage. Further losses occurred in June and July 1944 in central Germany, so that the 100 temporarily available He 177s shrank in number again alarmingly.

Other than the losses, the operational debut of the He 177-equipped KG 1, planned for early summer 1944, was again delayed due to the necessary reconfiguration and re-equipment of the He 177A-3s: the aircraft had to be altered at the Sagan-Küpper workshops from Equipment Condition 'B' (medium-range bombers) to Equipment Condition 'A' (short-range bombers). In the latter configuration the He 177 could carry considerably heavier bomb-loads over shorter distances. But once again planning and reality only met part of the way, due to the very short time allowed for preparations and the structural alterations that had to be completed in each aircraft in just four days. The situation was compounded still further by the fact that only 29 conversion sets were on hand at Brandis. As a consequence, *Oberstleutnant* Horst von Riesen's KG 1 needed several weeks to recuperate from losses suffered on 28 July 1944 before he could once again field 67 He 177A-3s, of which 56 were fully operational.

Most of these aircraft, together with a considerable number of new He 177A-5s from the Sagan-Küpper workshops and airfields in central Germany, were transferred to Prowehren and Seerappen in East Prussia. Previously, the *Stab* and II/KG 1 had flown their Ju 88As from these bases on sorties against advancing Soviet formations; but the poor condition of the airfields was very unfavourable for the forthcoming operations by the larger and heavier He 177, and there followed a series of accidents involving broken undercarriage units. The hangar capacity too seemed completely insufficient, and the same applied to the fuel and bomb stocks. The necessary aviation fuel was supposed to be delivered by the *Deutsche Reichsbahn* directly to the airfield apron, but due to numerous low-level attacks the supplies often arrived only after some delay.

The first three attacks by I and II/KG 1 He 177s were carried out in level flight without any losses, and with good results. The strongest raid was by 87 He 177s against railway targets in the Velikiye Luki area, about 450 km (280 miles) west of Moscow. The participating *Staffeln* flew in three large 'attack wedges' of about 30 aircraft each, loaded with four SC 250 bombs apiece.

When, on 23 June 1944, strong Soviet forces broke through the German front in the central sector of the Eastern Front, *Reichsmarschall* Göring ordered KG 1 to attack the enemy armoured

The aircraft of Geschwaderkommodore Oberst von Riesen, who led 87 He 177s a successful att on railway targ in the Velikiye area of Russia June 1944.

This He 177A-3 (V4+IN/Wk-Nr 332610) of 5./KG 1 was unserviceable for a few days after suffering engine failure during an operation on 28 June 1944.

The Start of Operational Service

Another operational He 177A-3 assigned to I/KG 1 for operations on the Eastern Front in summer 1944.

formations at very low level. When *Oberstleutnant* von Riesen questioned the suitability of the He 177 for this task, Göring pointed out to him that the Western Allies had successfully used 'carpet bombing' during the course of a tank battle. In doing so, Göring displayed a complete lack of insight, yet insisted that his orders be carried out.

The first such low-level operation was to be flown by all 24 serviceable He 177s. The aircraft, flying in pairs to cover each other, would attack Soviet Army tanks. But such tactics were useless when countered by effective anti-aircraft fire, and during the first attack KG 1 lost more than 10 of its aircraft. Despite the losses, even more costly low-level attacks followed a short while later.

On 20 July 1944 KG 1 received orders for a maximum effort to stop the Soviet tanks, as

Above and overleaf (top):
This He 177A-3 was one of the aircraft that participated in the very low-level attacks on Soviet Army tanks in July 1944. Heavy losses to anti-aircraft fire during these raids led to the withdrawal from operations of I and II/KG 1's He 177s on 28 July 1944.

An unidentified KG 1 crew shortly before another operational flight against targets on the Eastern Front in summer 1944.

always. During another low-level offensive sortie two He 177s had to abandon their bombing runs whilst approaching the target area. Altogether, the Soviet Army's mobile anti-aircraft defences and the numerous infantry weapons exacted a high price from the large, unarmoured He 177s attacking at low level. For this reason, I and II/KG 1 were withdrawn from operations by 28 July 1944. The surviving, and in some cases damaged He 177s, arrived back at various airfields in central Germany a short while later. There, some of the A-3s were dismantled and their powerplants mothballed, whilst other aircraft were reduced to scrap.

III/KG 1, operating as an independent unit since April 1943, was partly equipped with Ju 88 bombers and deployed against Soviet Army tanks on the Eastern Front in the sectors under the command of 1. and 2. *Fliegerdivisionen*. From May 1944 onwards, III/KG 1 started to re-equip with the He 177 at Burg bei Magdeburg. Before that, a few aircrews had already participated in He 177 operations against railway targets in the East.

KG 1 also had its own training and replacement unit, IV *Gruppe*, formed on 1 March 1944, although the first of its He 177A-1s did not arrive until the following month. These aircraft were at first used to train the necessary instructional personnel for I/KG 1. By 14 April 1944 IV/KG 1 had 22 A-1s and A-3s, all of which were unserviceable. The *Gruppe*'s personnel consisted of 14 instructor and 24 trainee crews, of which 10 were detailed to join I and II/KG 1 without any training on the He 177. After handing over the last of its He 177s, some of the personnel were transferred to other flying units, while others were detailed to the *Luftwaffenfelddivisionen* (Air Force Field Divisions).

Chapter 3

The He 177 in Maritime Warfare

From He 177A-3 to 'Spring Bomber'

During the *Führer* conference on 3 January 1943 Hitler admitted to *Reichsminister* Prof Albert Speer, Minister of Armaments and War Production, that considering its hasty activation, the demands made of the *Luftwaffe* had often been too great. A short while later, Hitler himself stated that:

> 'The He 177 was to be developed simultaneously as a four-engined heavy bomber and a dive-bomber. But I never thought anything of that! Only one of these attributes could be fulfilled, and because of that the entire development was drawn out uselessly for several years.'

Severe criticism also came from Dipl-Ing Heinrich Hertel, Heinkel's former Technical Director and Chief of Development, who had left the company in March 1939, only to return in late 1942 in his capacity as an RLM-appointed Commissar for the He 177 programme. In his opinion, there was no doubt that the all too frequent engine fires were due above all else to faulty installation of the powerplants at the Heinkel works. To the RLM he remarked that 'everything has been unlovingly thrown together and lacks any care and attention. In peacetime the aircraft would have been barred right away'.

Hertel's negative assessment was countered by the mostly positive experiences of I/FKG 50 crews during the Stalingrad support operations in January 1943. Qualified praise of the He 177 was expressed by the *Gruppenkommandeur*, *Hauptmann* Schlosser:

> 'On operations, the He 177 proved itself outstandingly in a tactical sense. However, its technical shortcomings, especially as regards the powerplants, not only considerably lowered the operational readiness, but also cost five aircraft and

...uably the ...kest element ...e He 177 was ...Daimler-Benz ...led ...erplants, the ...ntenance and ...icing of which ...igh demands ...he technical ...onnel.

the lives of four crews not due to enemy action during the first 14 days of operations. The elimination of technical shortcomings is an unconditional necessity.'

All this should in no way give the impression that all was well with the He 177; flight safety was and remained unpredictable on the large-scale series production aircraft. Indeed, there was still a long way to go to the standard production model, the He 177A-5, the so-called 'Spring Bomber'.

Meanwhile, Heinkel works and *Luftwaffe* testing had once again claimed new victims. On 2 February 1943 an A-3 (BL+FU/Wk-Nr 15191) crashed on its third flight whilst conducting stalling tests; the aircraft spun in from an altitude of about 4,000 m (13,120 ft) after one of its coupled powerplants unexpectedly failed, making it impossible for the pilot to hold the heavy aircraft aloft on the power of just one DB 606.

By then, concerted efforts were underway to purge the He 177 of its powerplant problems. Following Dipl-Ing Hertel's pleas, the RLM

Airfield fire service personnel in action at Châteaudun. Given its propensity to burst into flames, it is perhaps appropriate that the aircraft's engine is receiving their attention!

ordered that one A-3 from the Arado licence-production line (ND+SS/Wk-Nr 135024) be held back and fitted with another, improved version of the DB 610 coupled powerplant as soon as possible.

As a preliminary measure, the Heinkel works had been instructed to build a corresponding powerplant mock-up. Work progressed apace, and the pattern aircraft began flight-testing in March 1943. To save time, a comprehensive work programme had been agreed between the RLM and Prof Dr-Ing Heinkel on 16 February 1943:

> 'The He 177 powerplants will be fundamentally put in order by moving them forward by 200 mm [7.87 in], improvement of the wiring layout, provision of exhaust gas disposal, the installation of fire extinguishers, and the clearing up of other outstanding complaints.'

The first experimental carrier of the moved-forward DB 610s was to be the He 177A-3/V23 (VF+QL/Wk-Nr 332112), scheduled for completion by 1 March 1943. By 27 February the Heinkel works had also managed to complete the fitting of a mock-up fireproof powerplant installation on the He 177A-09/V10 (GA+QR/Wk-Nr 00 0024).

The first pattern aircraft with completely modified and repositioned DB 610s was followed by the He 177A-3/V19 and, a little later, another two experimental aircraft. Next to the urgent elimination of all powerplant problems, the staff at the Heinkel works were simultaneously working on the following pattern aircraft, some of which promised better flying characteristics:

He 177 V5 (PM+OD/Wk-Nr 00 0005):
Rebuild of series pattern elevators
He 177A-08/V9 (GA+QQ/Wk-Nr 00 0023):
Change from DB 606s to DB 610s
He 177A-3/V15 (Wk-Nr 335001):
Fitting of power-operated servo-tabs in ailerons
He 177A-3/V19 (VF+QA/Wk-Nr 332101):
Fitting of a new slotted wing
He 177A-3/V20 (VF+RD/Wk-Nr 15254):
Rebuild with new fuel feed system and fire extinguisher installations
He 177A-3/V21 (VF+QB/Wk-Nr 332102):
Modification as experimental carrier of *Kehl IV* guided bomb control device
He 177A-3/V22 (VF+QD/Wk-Nr 332104):
Fitting of a new slotted wing
He 177A-3/V24 (ND+SS/Wk-Nr 335004):
Powerplant installation
He 177A-3/V27 (V4+UC/Wk-Nr 15203):
Enlargement of rudder and modification of servo-tabs

Before the He 177 could undertake long-range maritime flights, it was necessary to ensure operational safety. This engine test-bed unit helped achieve that goal by conducting rigorous tests of the powerplants.

He 177A-3 (DH+CY/Wk-Nr 15199):
Powerplant performance measurement trials
He 177A-3 (VF+RC/Wk-Nr 15253):
Modification of flying controls

As can be seen from the accompanying table, testing of the He 177A-3 was progressing at full speed. The main emphasis was on fitting the more powerful DB 610 coupled powerplants and, of equal importance, strengthening the A-3's defensive armament. The planned armament comprised the following:

- Nose (upper): 1 x MG 81Z (manually-operated)
- Nose (lower): 1 x MG 151 (manually-operated)
- Dorsal (forward): 1 x MG 131Z (remote-controlled)

- Dorsal (aft): 1 x MG 131 (manually-operated)
- Tail: 1 x MG 151/20 or MG 131 (manually-operated)

The onboard radio equipment too was improved and consisted of the FuG 10P, FuG 17Z, FuG 25a, FuG 101, FuG 102, FuG 200, FuG 216 (FuG 203) and FuBl 2F. However, most He 177A-3s did not have all of this equipment installed.

The main difference between the A-3 and the A-1 was in the introduction of the Standard Equipment Set 'E': the *Kehl* 'special weapon' guidance installation enabling the A-3 to carry various guided bombs/missiles.

On 18 February 1943, following a request from the RLM, the Heinkel works prepared a precise tabulation of the sub-variants evolved from the He 177A-3, as follows:

models would differ only by virtue of the requested strengthened defensive armament. As for the repositioned DB 610 powerplants, the design bureau emphasised that they should be considered as a standard feature on all He 177s by the A-5 model at the latest.

So much for the efforts of the Heinkel works to ready their long-range bombers for offensive operations as quickly as possible.

In his five-part 'Advisory Comments' Prof Dipl-Ing Hertel covered all aspects of the He 177's development to date, and once again was forthright in his criticism and suggestions. In his opinion, it would not be advisable to change over to the four-engined He 177B (*see Chapter 4*) at short notice. Better instead to invest in the development of a long-range bomber with an all-up weight of 34,000 kg (74,956 lb), a longer wing span and the elongated fuselage of the A-3/A-5 models.

Projected Development Stages of the He 177

	He 177A-5	He 177A-6	He 177A-7
Features:	Longer fuselage Repositioned engines Full-strength wing	Longer fuselage Repositioned engines	New cockpit layout Repositioned engines New centre of gravity
Armament changes:		MG 131I (nose) MG 131Z (ventral) MG 131I (tail)	MG 151Z (nose) MG 151Z (dorsal) MG 131V (tail)
Delivery of construction data:	31 May 1943	30 June 1943	15 September 1943
Construction of two prototypes:	1 September 1943	1 December 1943	June 1944
Delivery of first production series aircraft:	Autumn 1943	Spring 1944	Mid-1945

After the realisation that the planned He 177A-4 high-altitude bomber would not reach fruition in the short term, development of the A-5, the so-called 'Spring Bomber' was allocated the highest priority. This long-range model was a direct development of the A-3 which, apart from new ailerons, a longer fuselage and in a 'special action', powerplants moved forward by 20 cm (7.87 in), corresponded to its predecessor.

As a result of the changes incorporated in the A-5, Prof Dr-Ing Heinkel claimed that the worst faults of the He 177 had been eliminated once and for all. In addition, the Heinkel design bureau insisted that all surviving He 177s of those already produced should receive the elongated fuselage, so that the next two production series

Hertel did not believe that the He 177B offered a notable improvement either in load-carrying capacity or engine power; while the large number of new components necessary for He 177B series production was seen as another problem.

According to figures supplied by JFM, their version of the He 177, in addition to being considerably easier to manufacture, would have an operational range of 5,200 km (3,231 miles) – 390 km (242 miles) and 750 km (466 miles) better than the He 177A-3 and He 177B respectively.

While JFM touted its version of the He 177, the staff at Heinkel's Rostock-Marienehe works were just as active. For his part, Prof Dipl-Ing Siegfried Günter prepared a comprehensive study detailing improvements to the He 177; it was

THE HE 177 IN MARITIME WARFARE   97

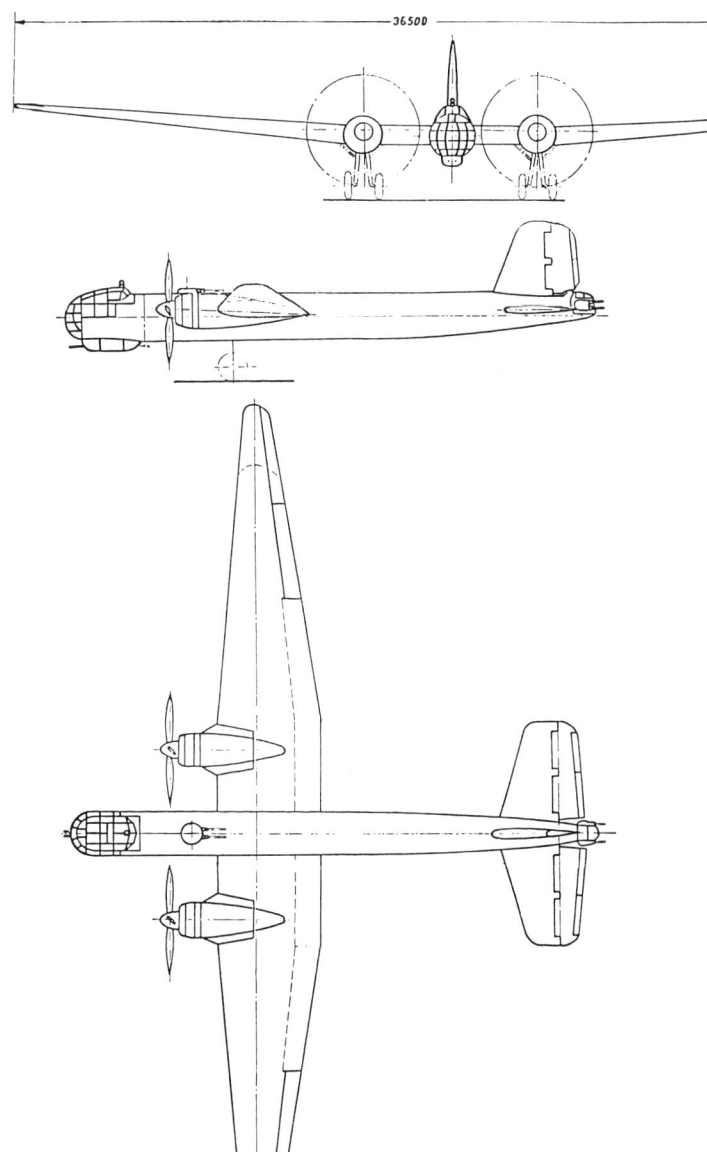

A three-view drawing of the 'Ju 177' powered by two DB 613s.

submitted to the RLM on 22 March 1943.

According to Dipl-Ing Günter, the best improvement in performance was promised by the installation of the heavier DB 613 coupled powerplant made up of two DB 603 12-cylinder liquid-cooled engines. However, keeping the existing main undercarriage units (but with bigger tyres) would require a thicker wing of deeper leading-edge section. Apart from the additional construction effort necessary, the new and heavier wing would alter the aerodynamics of the He 177. As a consequence, the increased all-up weight of the aircraft would necessitate strengthening of the fuselage, tailplane and tailwheel. Yet another drawback would be the marked loss of speed and a five per cent reduction in operational range caused by the fitment of four engine exhaust flame dampers.

For these reasons Prof Dipl-Ing Günter's second proposal envisaged the adoption of a new wing centre-section and constructive alteration of the outer wing sections. This would not only result in an optimum safeguard against engine fires but also enable the strengthened, larger main undercarriage units to be used without any problems. The drawback in Günter's second proposal was that the entire Fowler flap installation as well as the integral wing fuel tanks would have to be modified.

To avoid this, and contrary to Prof Dipl-Ing Hertel's belief, Günter proposed an immediate changeover to an aircraft powered by four individual air-cooled engines. Such a change would enable additional wing fuel tanks to be fitted, thus boosting range, and also considerably ease powerplant servicing and maintenance. At the same time Günter requested a change to a twin fin/rudder assembly as quickly as possible, and a corresponding redesign of the He 177's rear fuselage.

The fuselage as a whole was to be extensively modified. Suggestions by Major Mons on how to improve the existing cockpit to provide better

98 HEINKEL HE 177-277-274

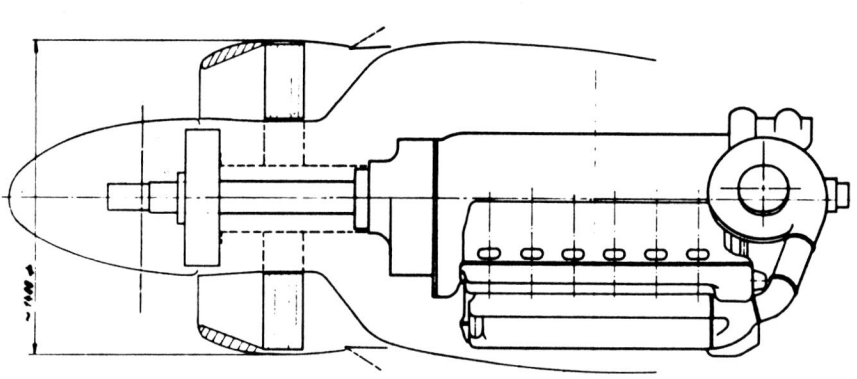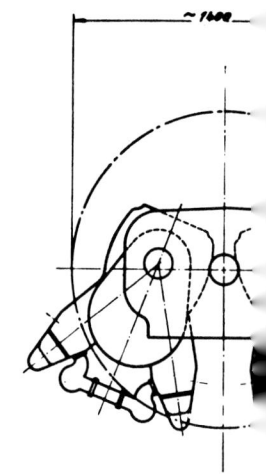

General outline drawing of the DB 613 coupled powerplant (two DB 603s) intended for the He 177.

vision were gratefully accepted by Heinkel and, in early 1943, led to the construction of two pattern cockpits which were later flight-tested. Another suggestion was to make the tail gun position accessible from the aft dorsal turret or the main cabin by means of a crawl gangway. Major Mons also proposed the transfer aft of the ventral gun position to a point behind the bomb-bay, in an attempt to provide a better field of fire.

Comparison of the proposals regarding the airframe and the future of the He 177A-5/A-7 with those submitted by Prof Dipl-Ing Günter revealed the following figures:

Comparison of Airframes

	A-6	A-7	177	177	177	177
Crew	5	5	5	5	5	7
Powerplant	2 ˇ DB 610	2 ˇ DB 610	2 ˇ DB 613	4 ˇ DB 603G	4 ˇ BMW 801E	4 ˇ BMW 801E
Output (each)	2,950 hp	2,950 hp	3,800 hp	1,900 hp	2,000 hp	2,000 hp
Wing Span	31.40 m	35.00 m	31.60 m	40.00 m	40.00 m	40.00 m
	(103 ft 0 in)	(114 ft 10 in)	(103 ft 8 in)	(131 ft 3 in)	(131 ft 3 in)	(131 ft 3 in)
Wing Area	102.00 m^2	107.00 m^2	108.00 m^2	133.00 m^2	133.00 m^2	133.00 m^2
	(1,098.00 ft^2)	(1,151.74 ft^2)	(1,162.50 ft^2)	(1,431.60 ft^2)	(1,431.60 ft^2)	(1,431.60 ft^2)
Weight (loaded)	32,000 kg	33,000 kg	34,700 kg	38,000 kg	42,000 kg	42,000 kg
	(70,548 lb)	(72,753 lb)	(76,500 lb)	(83,776 lb)	(92, 594 lb)	(92,594 lb)
Maximum Speed*	425 km/h	420 km/h	425 km/h	430 km/h	440 km/h	435 km/h
	(264 mph)	(261 mph)	(264 mph)	(267 mph)	(273 mph)	(270 mph)
Ceiling	6,800 m	6,800 m	8,000 m	7,500 m	7,900 m	7,900 m
	(22,310 ft)	(22,310 ft)	(26,247 ft)	(24,606 ft)	(25,919 ft)	(25, 919 ft)
Range	4,200 km	4,600 km	4,600 km	5,250 km	6,100 km	5,200 km
	(2,610 miles)	(2,858 miles)	(2,858 miles)	(3,262 miles)	(3,790 miles)	(3,231 miles)
Armament						
Nose (lower)	1 ˇ FDL 131Z	1 ˇ FDL 131Z	1 ˇ FDL 131Z	1 ˇ FDL 131Z	1 ˇ FDL 151Z	1 ˇ FDL 151Z
Dorsal (fwd)	1 ˇ FDL 151Z	1 ˇ FDL 151Z	1 ˇ FDL 151Z	1 ˇ FDL 151Z	1 ˇ FDL 151Z	1 ˇ FDL 151Z
Dorsal (aft)	—	—	—	—	—	1 ˇ HD 151
Ventral	1 ˇ WL 131Z	1 ˇ WL 131Z	1 ˇ WL 131Z	1 ˇ WL 131Z	1 ˇ FDL 151Z	1 ˇ FDL 151Z
Tail	1 ˇ HL 131V	1 ˇ HL 131V	1 ˇ HL 131V	1 ˇ HL 131V	1 ˇ HL 131V	1 ˇ HL 131V

* At 6,000 m (19,685 ft) altitude

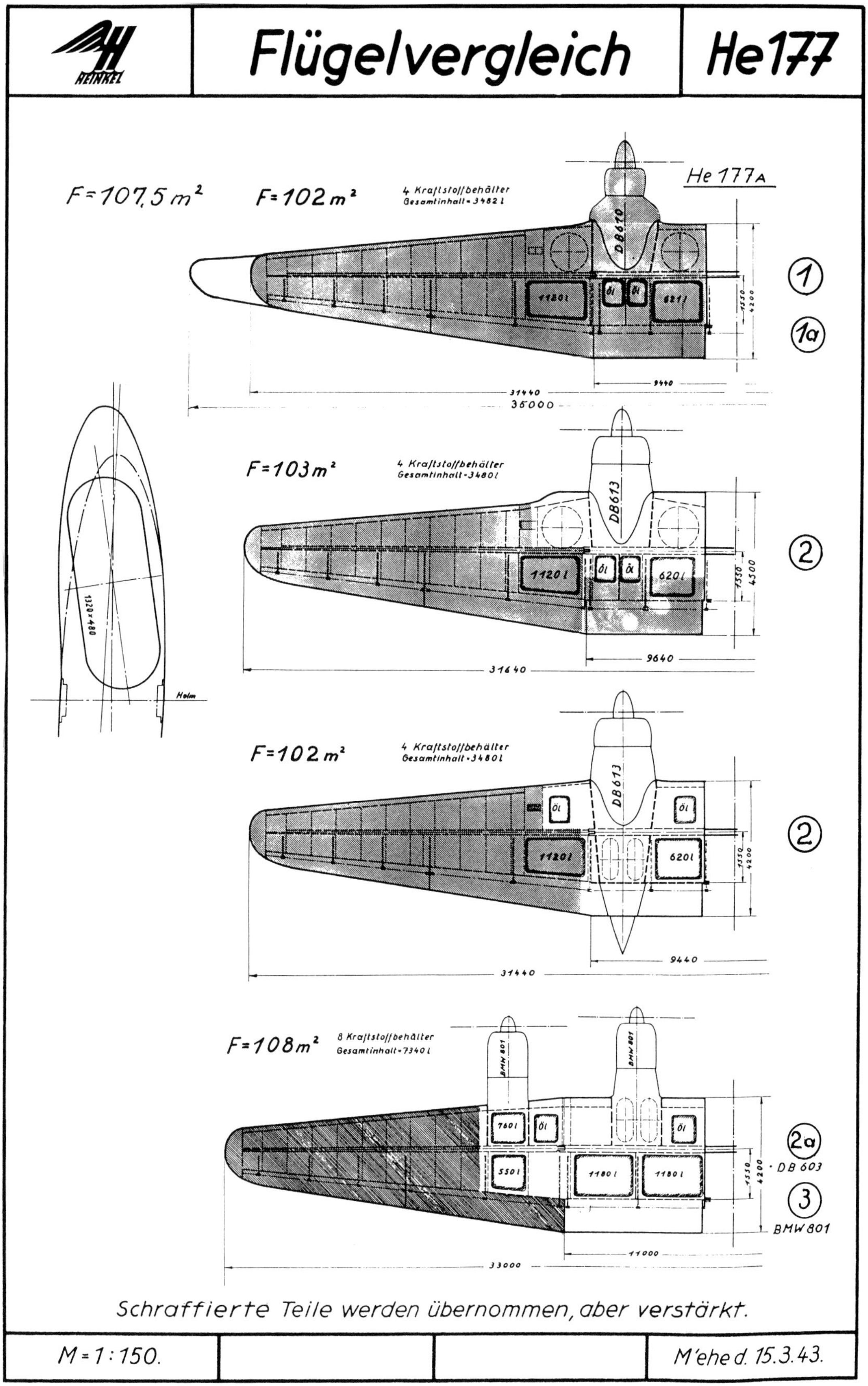

Above and overleaf:
A comparison of alternative two- and four-engined He 177 wings of increased span and area.

# Flügelvergleich — He 177

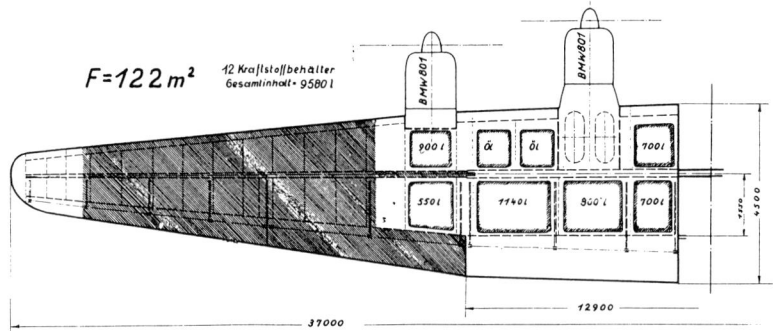

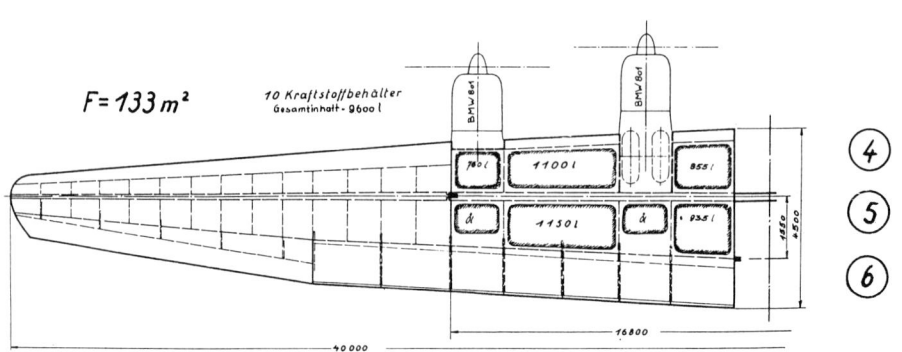

Schraffierte Teile werden übernommen, aber verstärkt.

M = 1 : 150.   M'ehe d. 15.3.43.

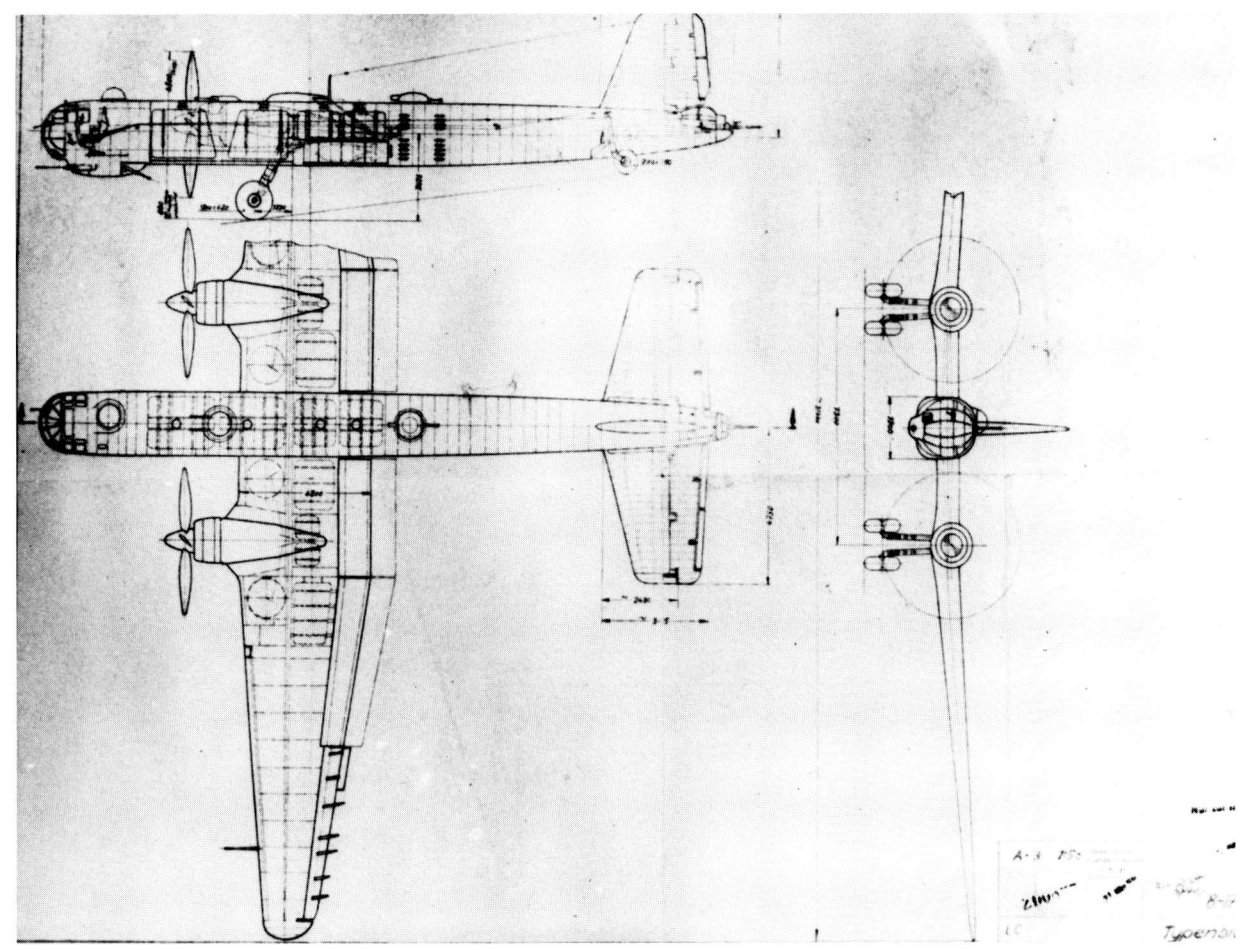

He 177 wing comparison up to 133.00 m≈ (1,431.60 ft≈).

The evaluation of all these data led in the end to a two-stage solution regarding planned further development:

a) He 177 with new wings and four separate DB 603 or BMW 801 engines, twin fin/rudder assembly and slightly longer fuselage.
b) He 177 similar to above but with new fuselage and cabin layout, reduction in crew to five men and deletion of aft dorsal turret.

Until matters had advanced that far, a '32-ton' bomber, the He 177A-6, was to be built as an interim solution. Work on this aircraft began in January 1943. In addition to the already-mentioned strengthened defensive armament, from the outset the A-6 was to feature the longer fuselage.

The first three examples of the 31,600 kg (69,665 lb) all-up weight A-6 long-range bomber were to be rebuilt and flight-tested at Rostock-Marienehe, then transferred to *E-Stelle* Tarnewitz for armament tests. The Heinkel works expected the aircraft to be ready by the end of 1943, but in fact the A-6 series was soon to be cancelled in favour of the He 177A-5 and B-5.

As far as testing of the repositioned DB 610 powerplants was concerned, the He 177A-3/V19 had already conducted its maiden flight. But during subsequent flight-testing in early April 1943 the crew encountered frequent overheating of the engine oil coolers. Needless to say, the

 **BEWAFFNUNG** — **He 177**

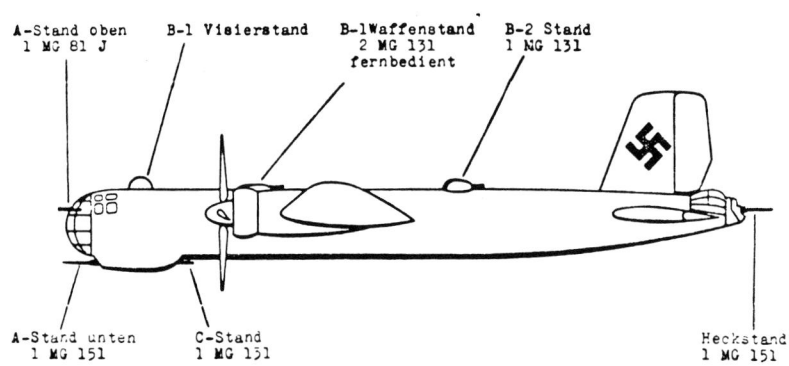

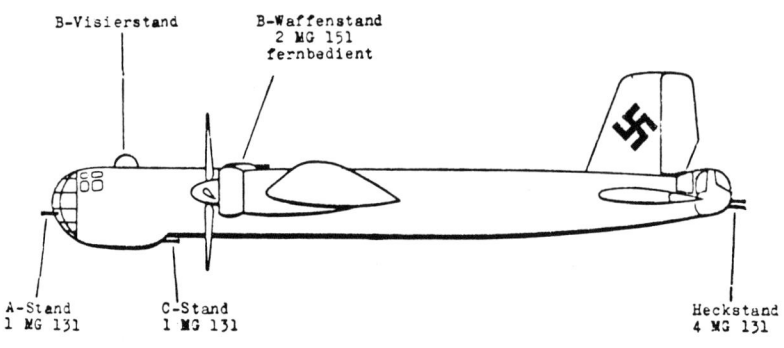

Defensive armament data for the He 177 A-3, A-5 and A-6 models. The most obvious difference is the deletion on the A-6 of the aft dorsal turret.

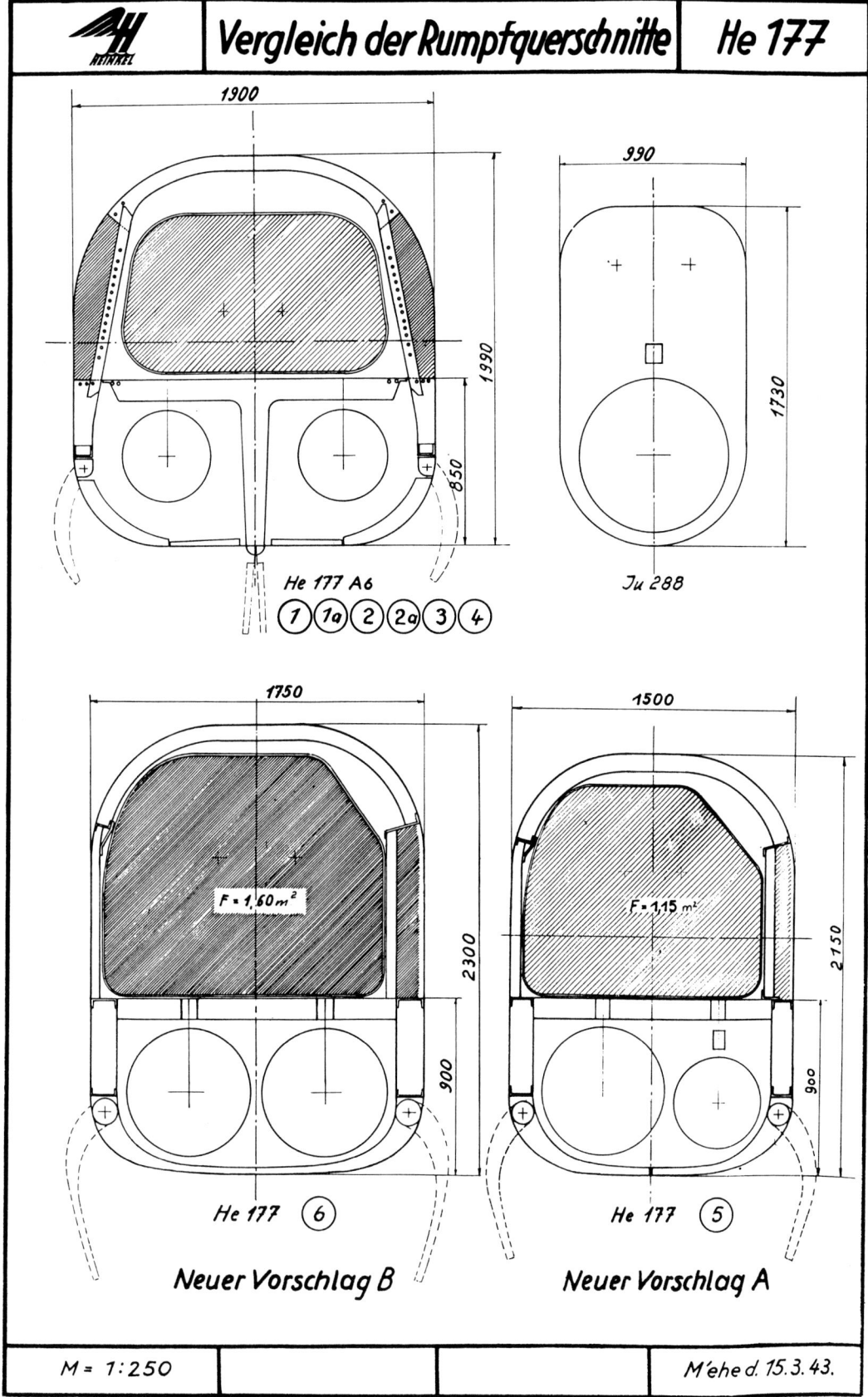

A comparison of three He 177 fuselage cross-sections with that of the much narrower Ju 288.

shortfall in deliveries of engines by Daimler-Benz and problems with oil leaks created additional obstacles for the engine test programme.

The first production series aircraft allocated for powerplant tests, the He 177A-3/V22, arrived at *E-Stelle* Rechlin in May 1943, followed a short while later by the A-3/V20 and A-3/V23. Apart from powerplant tests, the following tasks were also to be completed at Vienna-Schwechat and Zwölfaxing:

- Flight-testing of 15 experimental aircraft
- Further development of the wings
- Construction of three He 177A-7 prototypes
- Progressive development of the He 177
- Construction of three He 277 prototypes
- Construction of 10 He 277A-0 pre-production series aircraft
- Progressive development of the He 277

That the engine tests had not yet made the vital breakthrough and identified a reliable solution became evident during a visit to *E-Stelle* Rechlin by General Vorwald (GL/C), *Oberst* Petersen (KdE) and *Oberst* von Lossberg on 27 May 1943. General Vorwald sharply criticised 'the complete failure of the Daimler-Benz firm'; the first He 177 with repositioned engines 'represented nothing but an unbearable life of suffering'.

The DB 610s had to be changed three times between 22 March and 3 May 1943. Altogether the He 177A-3/V19 only managed to fly for about 10 hours. Worse was to follow: during a conversation with *Generalfeldmarschall* Milch in June 1943, *Reichsmarschall* Göring once again expressed his doubts regarding the operational readiness of the He 177. Reports had reached him from the experimental establishments stating that fitted with engine exhaust flame dampers, the He 177 just about managed to reach a top speed of 385 km/h (239 mph).

Meanwhile, feverish activity was underway in the Vienna area to improve the flight safety of the He 177; while a start had been made on the He 177A-7, a project offer having been prepared and submitted. But the dice had already been thrown in the RLM and the planned large-scale production of the A-7 was cancelled.

The He 177A-5, production of which was scheduled to start in spring 1944, was, like the A-7, now seen by the RLM as merely an interim solution pending development of an He 177 long-range bomber powered by four separate engines. The A-5 would have the same bomb-release, powerplant and fuel feed systems as the He 177A-3, but would feature a different frontal defensive armament configuration as well as the ability to carry aerial torpedoes as per Standard Equipment Set 'D'. Nevertheless, the A-5's primary role remained that of carrier for various controllable (i.e. guided) weapons. Production of the A-5 totalled at least 565 airframes in six sub-

During a visit by the local Gauleiter (District Leader of the National Socialist Party) to the Eger plant on 8 June 1944, he was shown around an He 177A-5. Note the underwing weapon rack.

models (A-5/R1-R6 heavy bombers, with the R7 heavy bomber a project study only).

The He 177 as a Torpedo-Bomber

In May 1940 a fuselage mock-up of the He 177 at Rostock-Marienehe was experimentally loaded

An He 177A-5 wearing night-black camouflage on its fuselage and undersurfaces.

with an LT 5W aerial torpedo minus its rudder; work on a newly-designed tailplane to fit the dimensions of the He 177's bomb-bay was already underway. The loading trials were the result of a proposal early in the He 177 development programme that required the He 177A-1 to be able to accommodate two aerial torpedoes internally. It was also intended that the angled shot, depth rudder and starter setting would be electrically adjusted. However, because of the rather intricate and expensive installation deemed necessary, the RLM requested that the torpedoes be carried externally instead.

Heinkel's first proposals to meet the request found immediate approval in the RLM, and the firm was quickly authorised to commence design work. A contract to build a pattern aircraft followed a few weeks later.

At first, only one aircraft was available for trials with aerial torpedoes, namely the He 177A-3/V17 (Wk-Nr 335005) at *E-Stelle* Travemünde. The tests began in summer 1942 and were later continued at *E-Stelle* Gotenhafen-Hexengrund* on Poland's Baltic coast. It was not until September 1942 that Heinkel equipped another pattern aircraft for further torpedo trials, this being the He 177A-3/V30 (ND+SM/Wk-Nr

* *Gdynia before and after the Second World War*

This He 177A-3 served with E-Stelle Gotenhafen-Hexengrund on torpedo trials for several months, and is seen here carrying the maximum load of four aerial torpedoes.

135018), which had been fitted with strengthened wings for heavier underwing loads.

On 16 October 1942 *E-Staffel* 177 formulated an opinion about aerial torpedo operations, but did so without actually having flown a single torpedo-dropping test flight! In his official report, *Gruppenkommandeur* Major Mons expressed the view that it would be somewhat difficult to fly such a large aircraft as the He 177 at very low altitude and for the length of time necessary to launch the torpedoes without significantly increasing the risk of the aircraft hitting the water.

In March 1943, an He 177A-3 was modified in the Lufthansa workshops at Berlin-Staaken to carry four aerial torpedoes externally (two beneath the fuselage; one beneath each wing). It was also intended to similarly modify the He 177A-04/V26 (GA+QM/Wk-Nr 00 0019). At that time, there was already a binding contract in force for the monthly output of 20 He 177A-3s with aerial torpedo attachments. According to instructions from the *Luftwaffe Generalquartiermeister* (6. Abt), production of these aircraft (designated A-3/R7s) was to start in June 1943.

In addition to the A-3s used to conduct torpedo-carrying and dropping trials, the He 177A-5/V31 was employed for static and flight trials with the Blohm und Voss LT 950 winged aerial torpedo (otherwise known as the L 10 or LT 10) *Friedensengel* (Angel of Peace). As a result of these trials, on 8 June 1944, two days after Allied forces stormed ashore in Normandy as part of the D-Day landings, the *Technische Amt* proposed that 25 He 177A-5s be configured for the carriage of two or three L 10s each. But this plan, like the one to manufacture the A-3/R7 torpedo-bomber, was scrapped some weeks later, due to the worsening military situation.

The He 177 *Zerstörer*

For 'special operations', the RLM requested the fitting of one or two fixed 30 mm MK 101 cannon in the He 177A-1's lower nose position. As a result, preparations were put in hand for the re-equipment of a small series of aircraft under the designation 'He 177 *Zerstörer*'. A further change to the armament configuration followed as the result of a request in February 1941 to fit four semi-fixed MG 151/20 cannon; a problem that had been largely resolved by the end of that year.

These *Zerstörer* aircraft were to be used to combat enemy reconnaissance aircraft and bombers operating over the Atlantic, which could pose a danger to the *Kriegsmarine*'s U-boats hunting Allied supply convoys. The aircraft were also intended to attack and shoot down Allied

The He 177 in Maritime Warfare   107

Above and below:
The He 177A-5/V31 (TM+IF/Wk-Nr 550202) was used for trials with two L 10 Friedensengel aerial torpedoes. The aircraft was also used for loading experiments with radio-guided 'special weapons'.

Close-up views of the L 10 aerial torpedo, showing its gliding wings and tailplane.

An L 10 is released from the He 177A-5/V31 during test-drops.

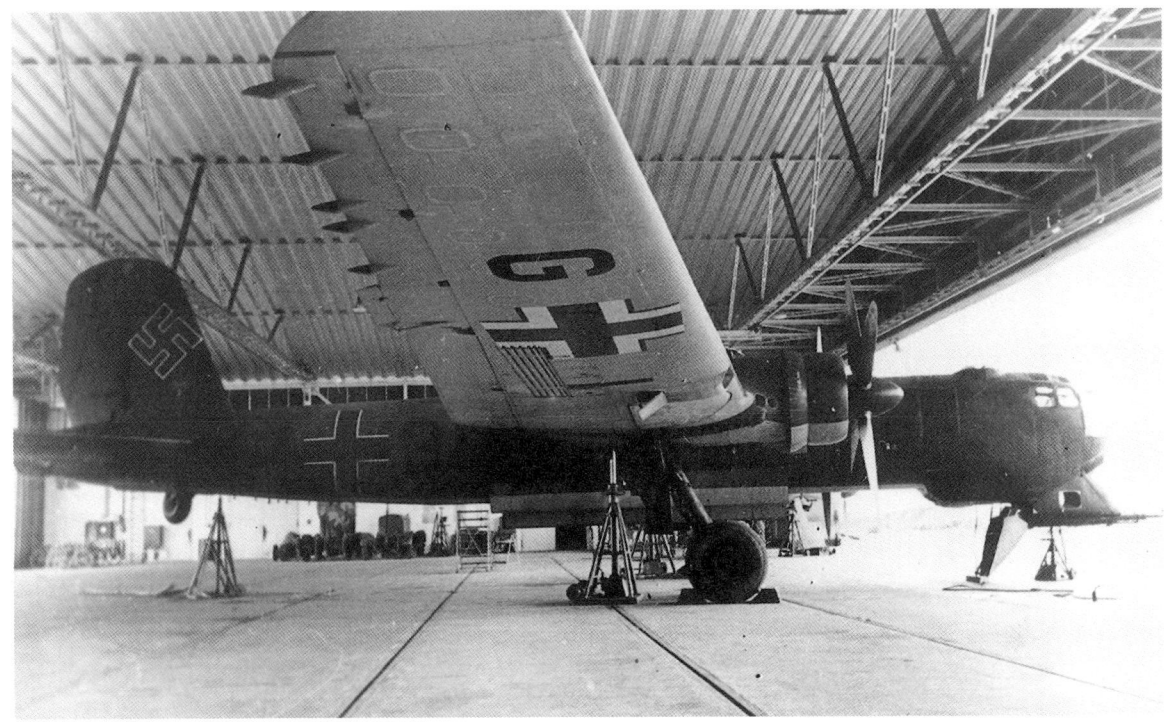

After the pattern installation of the 30 mm MK 101 cannon was complete, the He 177A-1/V12 (GI+BL/Wk-Nr 15151) was jacked up to check the functioning of its undercarriage.

four-engined bombers flying without fighter escort across the Atlantic Ocean to Great Britain.

To begin with, the necessary calculations were carried out at Rostock-Marienehe culminating in the construction of a wooden mock-up. Official approval of the project finally came from the RLM on 4 March 1942, enabling the rearmament of 12 He 177A-1 bombers to start at Rostock-Marienehe on 1 June 1942. Shortly thereafter, work began on the manufacture of the necessary conversion components and then the construction of a Standard Equipment Set comprising two 30 mm MK 101 cannon for the aircraft, now designated the He 177A-1/U2.

To achieve operational readiness as soon as possible, the first four A-1/U2s were supposed to be fitted with modernised powerplants. They were then to be transferred to *E-Staffel* 177 at Lärz early in June 1942, but due to work overload there it was decided to entrust the fitting of the new weapon conversion sets to I/FKG 50.

Then, quite unexpectedly, the RLM suddenly gave orders on 13 August 1942 to immediately halt construction of the He 177A-1/U2s and not to produce any more standard weapon equipment sets. The order followed a report by Major Schede of I/FKG 50 based on practical trials, according to which the additional armament was responsible for a loss of speed in the order of 100 km/h (62 mph)! However, comparative flights carried out later at *E-Stelle* Rechlin (E-5) resulted in a speed loss of only about 10 km/h (6.2 mph), and controlled tests by I/FKG 50 gave the same values. Clearly the initial test data were the result of inaccurate measurements, which in turn made the RLM's decision to halt further work on the aircraft all the more difficult to understand; especially considering that the original order for 12 such aircraft had been considered extremely urgent.

The first seven He 177A-1/U2s had to be delivered to the BAL by 25 September 1942, followed by the eighth, ninth and tenth aircraft from Heinkel early in October. On 15 October 1942 the first three examples stood ready for ferrying at Rostock-Marienehe. The following day, eight He

The first He 177A-1/U2 Zerstörer *pattern aircraft then underwent trials under operational conditions at* E-Stelle Tarnewitz.

177A-1/U2 production series aircraft were ready for collection. At that time, a further three *Zerstörer* aircraft were still with I/FKG 50 for flight test purposes. On 16 October, both MK 101 cannon were test-fired on GI+BR, while GI+BZ was delivered to the *Luftwaffe* a day later. With that, the order had been fulfilled by Heinkel ahead of schedule.

On 28 June 1943, three A-1/U2s (GI+BX, GI+BZ and Wk-Nr 15170) were with I/FKG 50 at Ludwigslust airfield, along with the He 177A-015 (GA+QX/Wk-Nr 00 0030), the second pattern aircraft which, after rebuild as a *Zerstörer*, had earlier been tested at *E-Stelle* Rechlin. These four aircraft were joined by four more A-1/U2s (GI+BQ, GI+BU, Wk-Nr 15169 and BL+FI). Another operational aircraft was at the Eger repair depot and, for a short while, at *E-Stelle*

After cancellation of the Zerstörer *programme, most of the He 177A-1/U2s stood around on various airfields for some considerable time before they were retrofitted to their original bomber configuration.*

An alternative Zerstörer configuration involved the fitment of two semi-fixed MK 101 cannon in the lower nose position.

Rechlin, but BL+FJ had crashed earlier in the year. It remains uncertain whether any of these aircraft were ever actually used for 'train-busting' purposes as mentioned by *Generalfeldmarschall* Milch during a conference on 11 June 1943.

Experiments with another special variant of the He 177A were conducted at *E-Stelle* Tarnewitz beginning around mid-December 1943. These concerned the use of built-in obliquely-angled Flak 43 (MK 103 cannon), the necessary structural data for which were hurriedly prepared by Rheinmettall-Borsig.

Another project, under consideration in summer 1943, sought to arm three He 177s as *Grosszerstörer* (Big Destroyers). To this end, each aircraft would be armed with 30 vertically-mounted infantry-type rocket launcher barrels in the modified bomb-bays. This proposal was subsequently discussed with *Generalfeldmarschall* Milch on 29 October 1943, when it was agreed to prepare the necessary plans. Construction of the first pattern aircraft took place at Diepholz during November 1943. By January 1944, five He 177A-3s had each been fitted with 33 vertically-mounted 21 cm (8/in) rocket launcher barrels, each angled to fire upwards at 60, at Quackenbrück and Diepholz. However one of the aircraft crashed and another three fell victim to an Allied low-level air raid on 21 February 1944, leaving only one *Grosszerstörer* for the time being. This aircraft was used at *E-Stelle* Tarnewitz to fire a total of 40 rocket shells on 23 and 25 March 1944.

On 11 April 1944 the same aircraft was demonstrated on the ground and in the air to *Generalleutnant* Galland, the *General der Jagdflieger* (GdJ: General of Fighters). Meanwhile, additional He 177s had been modified and, on 2 April 1944, all serviceable *Grosszerstörer* were transferred to Udetfeld in Upper Silesia for test-firing purposes. Later, three of the aircraft were transferred to *E-Stelle* Werneuchen to receive FuG 216/217 *Neptun* radar equipment, while another four were equipped with the FuG 16Y VHF transceiver.

In June 1944, all eight remaining *Grosszerstörer* aircraft arrived at Finow. After a second demonstration all 33 rocket shells on one aircraft reportedly fired of their own will (due to a faulty relay switch) right in front of the leading figures in the *Luftwaffe*. Whatever the technical fault, official faith in the *Grosszerstörer* was noticeably cooler after this event, and there are no details of any operational use of this variant.

The He 177 was also used for trials with 'target-released weapon-triggering' devices. The first of these tests took place without any significant success on 26 August 1944. By 30 September two specially-armed Fw 190A-8 fighters were ready for collection from *E-Stelle* Tarnewitz. The aircraft were from *Jagdgruppe* (JGr: Fighter Wing) 10, stationed at Tarnewitz at that time, and were each fitted with 18 SG 116 *Zellendusche* automatic target-released shells in 30 mm MK 103 cannon barrels brought up to the latest technical standard for the unit. The shells were fired in salvo by means of photo-electric cells located on the port side of the aircraft fuselage, these being activated by the shadow of the He 177 target aircraft.

By 9 October 1944, two so-armed Fw 190A-8s from JGr 10 had been made available for test-firing against an He 177 target aircraft. Both fighters were transferred to Parchin, and on 23 October the pilot of the first aircraft fired three rocket shells at an He 177. After two hits the third barrel suffered a misfire, while the complete electrical system short-circuited on the second Fw 190A-8. The relay switches were then improved, and the delay time of the shell release altered.

Both Fw 190A-8s were ready to resume trial attacks on an He 177 in mid-November. Several attempted test-firing flights failed due to faults in the electrical system aboard each fighter on 25 October and again at the end of November 1944, and new trials were postponed until the arrival of the AEG device. There were further problems with faulty photo-electric cells, but by 20 December 1944 the technicians had succeeded in equipping two Fw 190A-8s with a modified shell-firing installation. However, the practical experiments had to be temporarily suspended due to bad weather conditions.

By 19 February 1945, functional tests had been completed on all three aircraft, and flight trials could begin. One of the fighter pilots succeeded in achieving hits on the target He 177 during each of several mock attacks, but by early March

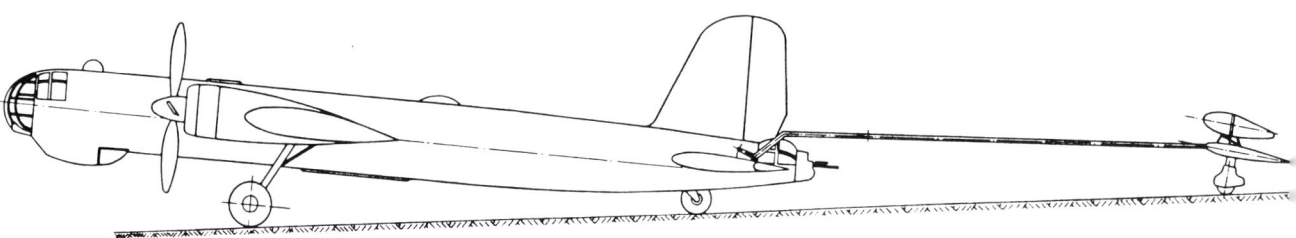

He 177A-5 with a Schleppgerät *(Towed Device).*

This He 177A-5 (TM+IU/Wk-Nr 550229) was one of two He 177s allocated to the DFS during the war. The aircraft's duties included towing the Schleppgerät *on a series of flight tests during June 1944.*

THE HE 177 IN MARITIME WARFARE 113

The roll-stabilised **Schleppgerät** *was designed to carry a 300-ltr (66 Imp gal) auxiliary fuel tank to help boost the He 177's operational range. Flight tests conducted during June 1944 evaluated the device's vulnerability to aerial gunfire.*

further technical problems and the deteriorating supply situation once again resulted in delays to this trials programme.

One He 177A-5 (TM+IU/Wk-Nr 550229) was also used as a towing aircraft for a roll-stabilised *Schleppgerät* (Towed Device) used by the *Deutsches Forschungsanstalt für Segelflug* (DFS: German Research Institute for Sailplanes) for gunfire hit examination purposes in summer 1944. Under the direction of Schieferdecker, the pilots Klöckner and Güttler carried out numerous flights in June 1944 at speeds of up to 620 km/h (385 mph). The *Schleppgerät* passed all tests, and was considered suitable by the DFS to carry up to 300 ltr (66 Imp gal) of additional fuel. Subsequent experiments with a 900 ltr (198 Imp gal) fuel carrier towed by the Me 262 and Ar 234 were based on these trials with the He 177A-5.

He 177 Trials with Guided Weapons

Development of the Hs 293 radio-controlled guided missile began in early 1939 under the direction of Prof Dr Herbert Wagner at Henschel Flugzeugwerke AG's Berlin-Schönefeld works. Based on the SC 500 general-purpose bomb, it sported light-alloy wings and tail, solenoid-driven ailerons and an electric screwjack driving the elevator.

Unfortunately, the long and comprehensive series of tests and experiments necessary required more time to complete than was originally expected, and so it was not until May 1940 that the first test device was released from an He 111. The first, still unguided Hs 293 V1 was dropped from an He 111H-4 (DC+CD) on 5 September 1940. By the end of the year, tests were underway using pre-production Hs 293A-0s fitted with an underslung pod for a Walter 109-507B rocket motor producing 600 kg (1,323 lb) of thrust for 10 seconds after ignition.

The structural data necessary for an Hs 293-carrying He 177 were delivered by 31 July 1941, and the construction of a pattern aircraft was authorised shortly thereafter. On the orders of the *Generalluftzeugmeister*, General Ernst Udet, 40 Do 217s and 40 He 177s were to be modified as quickly as possible for impending guided missile operations over the Atlantic Ocean. Of these aircraft, 34 of each type were to be equipped with the Hs 293 and six each with the PC 1400X (*Fritz-X*) radio-controlled glide-bomb. Simultaneously, endurance testing of the Hs 293A-1 was initiated by *Erprobungs und Lehr Kommando* (E-Lehr Kdo: Test and Instruction Detachment) 15. As the He 177 was not available on time, the number of Do 217s to be reconfigured as carriers was increased to 120 aircraft.

On 28 July 1942 *Generalfeldmarschall* Milch requested that the Hs 293, which would be launched from an altitude of at least 300 m (984 ft),

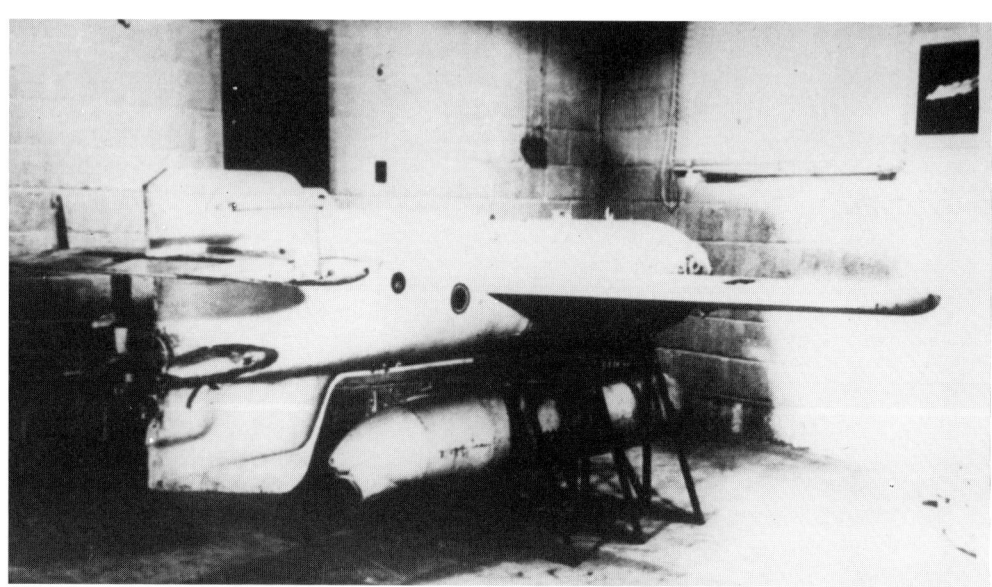

An Hs 293A-1 production series rocket-propelled radio-guided bomb.

Progressive development of the Hs 293 radio-guided rocket-propelled glide-bomb continued unflaggingly after delivery of the Hs 293A-1 production series.

should have the highest priority for the coming operations over the Atlantic. He also stated that he expected an 'all clear' report from *E-Lehr Kdo* 15 before the end of the year. But outstanding technical problems could not be solved in time, so the Hs 293's operational debut had to be postponed from month to month; an unsatisfactory state of affairs made worse by the Hs 293's initially very low serviceability rate. Also, the technical-tactical evaluation of the He 177 with underslung Hs 293s was still lacking, and there were many detailed questions that had yet to be answered.

A little later, following the crash of the He 177A-013 over the Plauer lake near Rechlin on 16 July 1942 (due to the fracture and subsequent loss of a wing), there was a surprising change in the programme: the first three He 177s scheduled to receive *Kehl* radio-guidance equipment for trials purposes were now detailed to undertake general flying tests instead.

One *Staffel* of I/KG 40 was earmarked for trials with the Hs 293, but as not enough He 177s were available, it meant yet more delays. On 13 May 1943, the following aircraft types were requested for operations with remote-controlled missiles and bombs: 24 He 111s, 100 Do 217E-5s, 50 Do 217K-2s, 34 He 177A-1s, 34 He 177A-3s and 40 He 177A-5s. Two months later, and in contrast to the Do 217E-5, the He 177 was still far from achieving operational readiness with a fully-functioning PC 1400X installation. Yet despite this, in August 1943 orders were received to re-equip the He 177A-3 model with *Kehl III* and *IV* sets for use in conjunction with the wire-guided Hs 293. Practical trials of the missile followed at *E-Stelle* Peenemünde using the He 177A-3/V21, the aircraft having been fitted with an ETC aft of the ventral gondola.

During the period from 3-28 October 1943, *E-Stelle* Karlshagen also determined the maximum range of an Hs 293-carrying He 177A-3 (GP+WY/Wk-Nr 535437), although the aircraft was only loaded with two such missiles. These controlled measurements recorded a speed of 270 km/h (168 mph) in climbing flight, and 380 km/h (236 mph) in level flight at an altitude of 3,000 m (9,840 ft), increasing to just 450 km/h (280 mph) at 5,500 m (18,045 ft).

Apart from everything else, the operational reliability of the *Kehl* radio-guidance equipment was of the utmost importance for the forthcoming operations over the Atlantic. The original trials with *Kehl I* and *II* sets were followed in June 1940

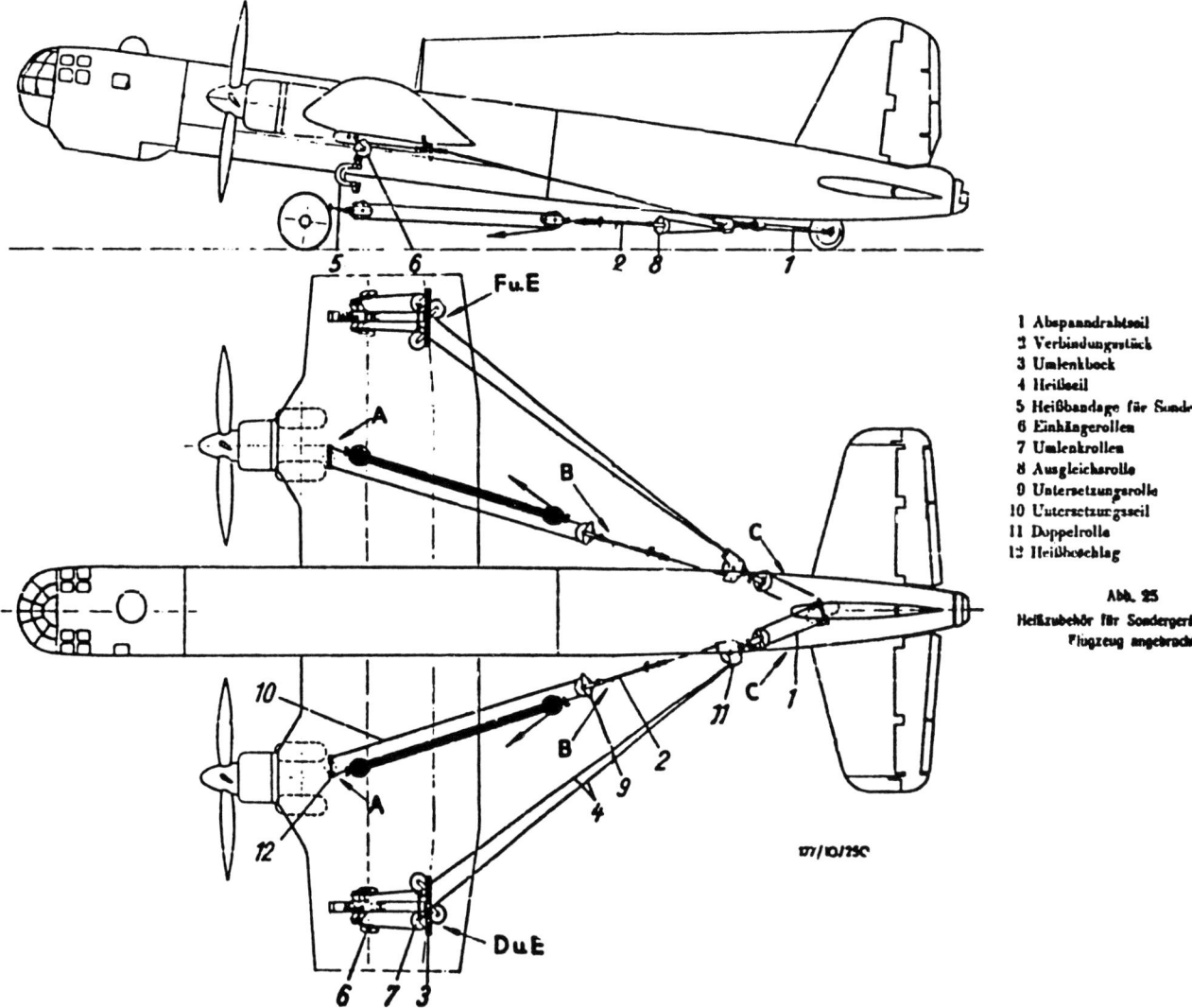

An excerpt from the He 177 'special weapons' handbook.

by testing of the improved *Kehl III*, but as only one He 111 was available to act as a test-bed for the new equipment, no quick success could be expected. Some experiments continued for many months before they achieved useful results. Eventually, there were six versions of the *Kehl* guidance set:

- *Kehl I* FuG 203a Remote control for one PC 1400X
- *Kehl II* FuG 203b Remote control for one Hs 293
- *Kehl III* FuG 203c Remote control for one PC 1400X or one Hs 293
- *Kehl IV* FuG 203d Remote control for up to four PC 1400Xs or Hs 293s
- *Kehl 1* FuG 203-1 Simplified device, nine channels
- *Kehl 2* FuG 203-2 Simplified device, nine channels

As the *Kehl III* installation was to be retrofitted on the He 177, it was essential to arrange a separate site where the re-equipment could take place without interfering with the current He 177 production programme. Several months later, after a protracted 'paper war', the choice fell on the airfield at Lüneburg, and Heinkel was allocated sufficient hangar space to enable the conversion work to take place. The company initially envisaged the attachment of one Hs 293 under each outer wing section, while the mock-up with the *Kehl III* installation prepared by *E-Lehr Kdo* 15 placed both missiles closer to the powerplants.

In April 1942, *E-Lehr Kdo* 15 delivered two modified He 177s to *E-Stelle* Peenemünde; both aircraft were still without the longer fuselage and improved, repositioned powerplants. During subsequent trials with the aircraft, it was established that heating of the 'special loads' was insufficient and needed urgent attention. The problem having been solved, in summer 1942 *Oberleutnant* Gold and *Feldwebel* Masuchni of II/KG 100 carried out the first successful glide-bomb releases using the *Kehl III* guidance equipment.

By 23 September the first six *Kehl III*-equipped He 177A-1s had been completed at Lüneburg, but delivery of the remaining 26 was delayed because of a lack of powerplants; as mentioned before, I/FKG 50 had no reserve engines. The first six aircraft reached operational units in mid-October 1942. The seventh A-1 earmarked for this task had also been modified in the meantime, but had to be diverted for a brief spell of inclined flight tests.

A total of 25 *Kehl III*-equipped He 177A-1s were at Lüneburg up to 23 April 1943, and in accordance with proposals made by the Heinkel works directorate, were duly converted *in situ* as *Kehl* training aircraft.

The well-rehearsed debut of the Hs 293 with II/KG 40 took place on 21 November 1943, but the operational use of what proved to be a troublesome stand-off missile lasted only a few months; the technical standard of the weapon was simply far too low. Another reason for the poor hit ratio was minimal visibility during attacks conducted at dusk; while insufficient experience on the part of the guidance operators, each of whom lay in the bomber's nose compartment and used a two-axes miniature joystick to try to guide a missile whose terminal velocity could reach 900 km/h (560 mph), undoubtedly played its part. In addition, Allied air superiority played a significant part in driving off the *Luftwaffe* attacks.

In addition to the Hs 293, several variants of which were produced, the He 177 was also to be armed with the Hs 294 remote-controlled bomb-torpedo. The first test-drop of this novel twin-rocket weapon took place from a Bf 110 on 7

Crouched in the nose section of an He 111, a bombardier works the control transmitter enabling line-of-sight guidance of an Hs 293A-1 radio-guided bomb.

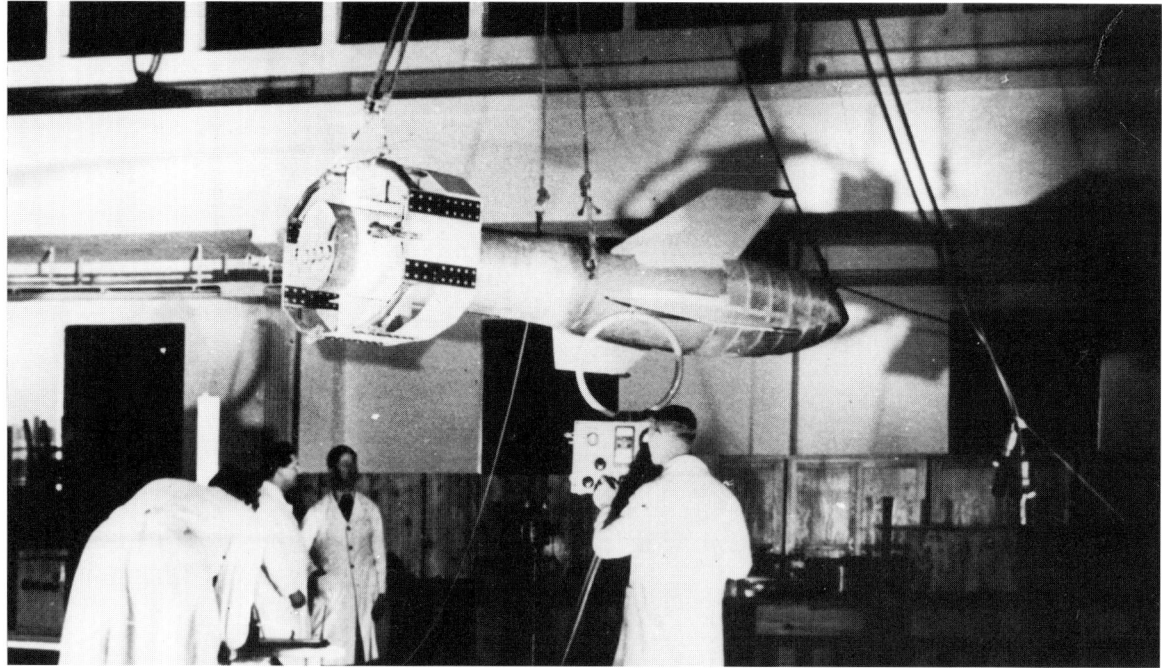

Trials with the PC 1400 Fritz-X radio-guided glide-bomb under laboratory conditions. Note the prominent 12-sided annular wing to restrict terminal velocity.

August 1942, but due to the state of its technical development further progress was rather slow. Experimental devices were built at Henschel's Berlin-Schönefeld plant and afterwards tried out on a Do 217K-2, but the He 177 had to wait for some time before its allocation was delivered.

By 27 April 1943, a total of 12 Hs 294s and an additional 15 water-running devices had been dropped by aircraft at *E-Stelle* Karlshagen; only seven of these drops were successful. Between 28 October and 1 November 1943, an He 177 dropped three Hs 294s without incident near Karlshagen, but that was all. Because of the Hs 293A missile and PC 1400X bomb, this new bomb-torpedo did not enter production, let alone operational use.

Extensive testing of the PC 1400X, basically a 1,400 kg (3,087 lb) armour-piercing glide-bomb conceived pre-war by Dr Max Kramer at the DVL as part of his X-series of guided weapons, had begun at *E-Stelle* Peenemünde in February 1942, and the dispersal pattern of the new weapon seemed extremely promising: the first drops were all within a circle of only 16 m (49 ft) diameter. Measuring 3.26 m (10 ft 8fi in) in length, the PC 1400X featured four fixed, angled wings and a cruciform tail, the latter surrounded by a 12-sided annular wing to restrict terminal velocity. The main control and radio compartment was towards the rear of a fuselage tipped by a 300 kg (661 lb) warhead in an armour-piercing casing.

Encouraged by the results of the initial trials, the *Luftwaffe* command activated *E-Lehr Kdo* 21 for the instruction and training of six guidance operators per week. But in November 1942 *E-Lehr Kdo* 21 was redesignated *KGrzbV* 21 and its Do 217s, He 111s and Ju 86s were pressed into the supply-dropping role in support of the beleaguered German 6th Army in Stalingrad. Not surprisingly, it took months before the full glide-bomb trials programme could be restarted.

The PC 1400X entered operational service on 29 August 1943, but the limited depth of penetration of the Do 217K-1/-2/-3s was considered insufficient from the outset. The He 177A-3 was deemed to be far more suitable for long-range operations over the Atlantic Ocean, and demand for a PC 1400X-carrying version of the Heinkel bomber received fresh impetus following the sinking of the Italian Navy's 46,485-tonne (45,732-ton) battleship *Roma* on 9 September 1943. The ship was sailing from La Spezia to Malta along with

An He 177A-3 after being fitted with an underfuselage weapon rack for a guided bomb at Fassberg in 1944. The aircraft was assigned to 8./KG 100.

two other battleships (the *Italia* and *Vittorio Veneto*), six cruisers and eight destroyers – the core of the Italian Fleet – in order to surrender to the Allies when they were intercepted and attacked in the Strait of Bonifacio by Do 217K-2s of II/KG 100. The *Roma* took two direct hits from PC 1400Xs, blew up and sank. The *Italia* was also hit but managed to reach Malta.

Once again, other factors intervened, and the He 177A-3 and A-5 remained equipped with Hs 293A-1/-2s; use of the PC 1400X was limited to the Do 217s of III/KG 100. Thus, during the final stages of its operational service III/KG 100 flew the Do 217M-11 armed with one PC 1400X semi-recessed beneath the fuselage.

To make the He 177 tactically more versatile, it was demanded that from July 1942 these aircraft should be able to carry either the Hs 293 or the PC 1400X, the optimum number being three of the former and up to four of the latter. A mock-up installation was built and examined on 11 August 1942 when, in addition to the PC 1400X attachment, the Hs 293/294 were also displayed and described. However, due to serious technical defects in the He 177's electrical circuitry, trials with the Hs 293 and *Kehl IV* guidance equipment were greatly hampered at the time.

The first experimental aircraft to act as prototype for the *Kehl IV* installation was the He 177A-3/V21, completed by the Heinkel works early in 1943. The aircraft was accepted by the BAL on 11 March 1943, and work soon began on extensive instrumentation tests. In April 1943 *E-Stelle* Peenemünde took delivery of the first production series He 177A-3 to feature the *Kehl IV* guidance system (VF+QC/Wk-Nr 332103).

There followed a series of loading experiments with the PC 1400X, which by that time had been fitted with mock-up wire spools. It soon became apparent that a PC 1400X attached beneath the He 177's fuselage lightly touched the main undercarriage units, so the idea of carrying two such bombs in tandem (originally suggested because the weapon had not yet been cleared for carriage beneath the He 177's wings, and everybody wanted quick results) had to be abandoned and a new solution found. *E-Stelle* Peenemünde suggested the use of only one underfuselage ETC. This would allow the He 177A-3 to carry the Hs 293/294/295 (the last of these being a twin-motor missile with an armour-piercing warhead), the *Fritz-X* or the *Peter-X* (a progressive development of the *Fritz-X*; development not completed) without any problems.

But there *were* problems. With the PC 1400X in place, the He 177A-3 had an all-up weight of approximately 31,000 kg (68,342 lb) and a very limited ground clearance beneath the bomb.

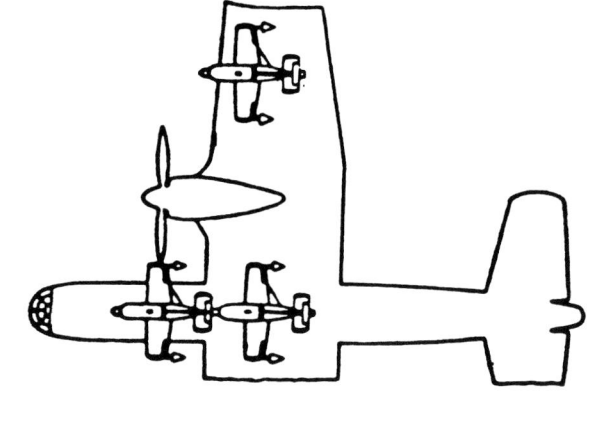

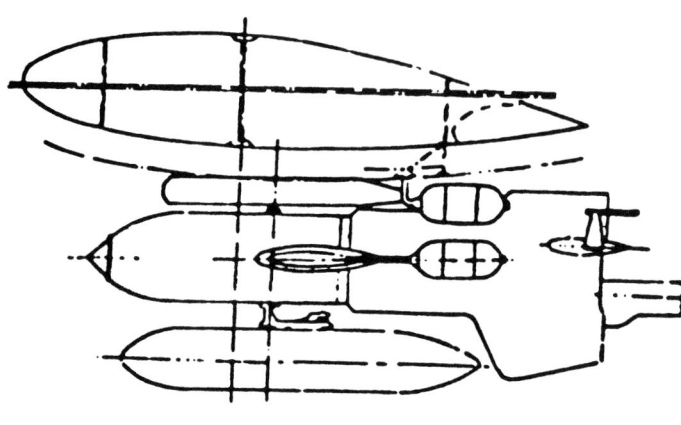

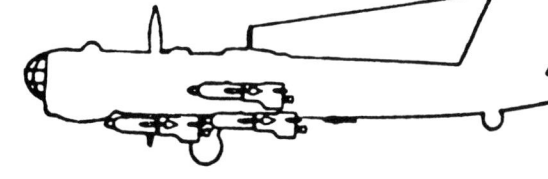

However, the experts at *E-Stelle* Peenemünde considered this acceptable, as the operational use of the aircraft would take place primarily from concrete runways and so there was no risk of damage to external drop-loads.

In an attempt to reduce the aircraft's all-up weight, *General* Vorwald (GL/C) suggested on 26 May 1943 that remote-controlled drop-loads be carried on the outboard underwing attachments only. The second pattern aircraft with *Kehl IV* equipment, supposed to be completed by Arado by 10 July 1943, was flight-tested in April, while the third *Kehl IV*-equipped test carrier was delivered on 15 May 1943. This aircraft featured only two ETCs, one under each outer wing section. In the event of an emergency, both could be blown off by explosive charges.

In autumn 1943, the reworked *Kehl IV* device was fitted into the specially-reserved He 177A-3/V36 (Wk-Nr 332121) and presented as the pattern installation for series production. It was approved without any problems soon afterwards. In the service manual for 'special weapons' it states:

> 'The attachment of special devices and drop-loads is achieved on two ETC 2000/XII electrically-operated racks equipped with Schloss 2000/XIIB locks, which are fitted under the port and starboard outer wing sections. The following kinds of ammunition can be loaded:
>
> 1) 2 x Hs 293   2) 2 x SC ammunition (e.g. Fritz-X)

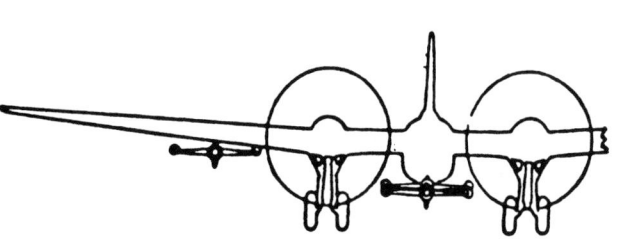

Schematic views of an He 177A-5 carrying underwing and underfuselage Hs 293A-1 radio-guided bombs, with a close-up of the underwing attachment.

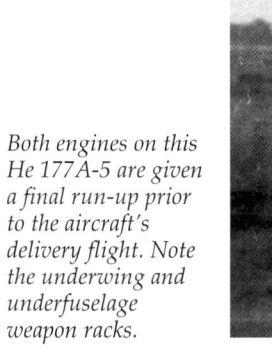

Both engines on this He 177A-5 are given a final run-up prior to the aircraft's delivery flight. Note the underwing and underfuselage weapon racks.

These drop-loads are released by means of the electrically-operated rack'.

It should be pointed out that the Hs 293 remote-controlled missiles were released via the *Kehl IV* equipment, while the PC 1400Xs were dropped by means of the normal bomb-release mechanism; there was no emergency jettison device. The drop-loads were armed by the He 177A-3/A-5s standard fuse-setting system.

The loading of these guided weapons was quite different from that of ordinary bombs. They were lifted into place by means of a special hoist and

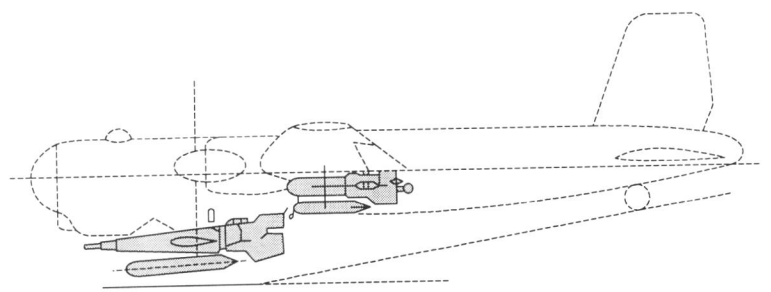

Attachment of two Hs 293A-1s (underwing) and one Hs 294 (underfuselage) radio-guided bombs on an He 177A-5.

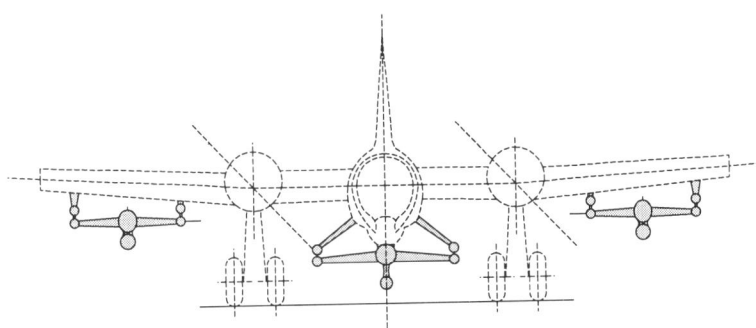

then attached by hand to the underwing ETCs. Both the weapons and their carrier racks were warmed by a heating device operated by hot engine gases. The finalised Equipment Condition 'E' was as follows:

necessary practical trials to *E-Lehr Kdo* 36. The first A-5 to be fitted with the *Kehl IV* was completed on 12 November 1943. By mid-December 10 additional aircraft had been likewise fitted at Arado's Brandenburg-Neuendorf plant.

Weapon Number	Type	Installation Fuselage	Wings	Attachment
3	Hs 293	1	2	Schloss 2000/XIIIB ETC 2000/XII D1
2	Hs 294	1	1	ETC 2000/XII D1
3	*Fritz-X*	1	2	ETC 2000/XII D1
3	*Peter-X*	1	2	ETC 2000/XII D1 ETC 2000/XII D1
2	Hs 293	–	2	ETC 2000/XII D1
1	Hs 294	1	–	ETC 2000/XII D1
2	Hs 293	–	2	ETC 2000/XII D1
1	*Peter-X*	1	–	ETC 2000/XII D1
2	*Fritz-X*	–	2	ETC 2000/XII D1

During later operations only the PC 1400X (*Fritz-X*) and HS 293 guided weapons were to be carried. Most of the aforementioned drop-loads were only attached on the racks for trial purposes.

According to the KdE, the *Kehl IV* equipment had to be tested on one of the first He 177A-5 production series aircraft; and he entrusted the

Most operational *Kehl IV*-equipped He 177A-5s allocated to II/KG 40 were rebuilt by *Feldwerft-Abt zbV* 1; the fuselage ETC was removed and the aircraft fitted with the *Kehl IV* equipment. However, examination of an A-5 (KM+US/Wk-Nr 550069) by the *Kehl* detachment revealed that the wire circuits and fuse protection were insuffi-

ciently or completely unprotected against dampness, which explained many of the previous equipment failures during operations.

As noted before, the *Kehl IV*-equipped He 177A-5 was intended to carry a single Hs 294 under the fuselage in addition to an Hs 293s under each outer wing section. Continuing doubts on the part of ground crews regarding insufficient ground clearance were thoroughly checked by *E-Stellen* Rechlin and Karlshagen in June 1944 and found to be unfounded.

Between 1 February and 30 June 1944, *E-Stelle* Karlshagen also determined the 'special weapon'-carrying He 177A-5's maximum range by using KM+TK/Wk-Nr 550035 to carry the following drop-loads:

- 1 ˜ Hs 293 (under fuselage)
- 1 ˜ PC 1400X (under fuselage; on ETC)
- 2 ˜ Hs 293 (under wings; on ETCs)
- 2 ˜ PC 1400X (under wings; on ETCs)

The flight procedure with all of the aforementioned loads was similar. At a take-of weight of 31,000 kg (68,342 lb) the aircraft would climb to a visual height of 300 m (985 ft) and then keep this altitude constant. Only just before the attack phase would the aircraft start to climb to the weapon-release altitude, using the maximum permissible constant power output in the process. Following a 30-min attack phase at an all-up weight of 27,000 kg (59,524 lb), the aircraft would bank away for a return flight at an average all-up weight of 24,000 kg (54,012 lb).

In reality, after the very costly operations on the Invasion Front in June 1944, it was clear that no further guided weapon sorties were likely to be flown in the West, and numerous trials already in progress thus became superfluous.

The He 177 in Service with KG 40

Activation of *Stab* and I/KG 40 began on 23 September 1939, both being equipped with Fw 200C Condors. Further expansion came during summer 1940 with the formation of the all-new II *Gruppe* (Do 217E) and III *Gruppe* (He 111H-6 and Fw 200C), formed from I/KG 1. These were followed in winter 1941-42 by IV (Erg) *Gruppe*, formed from KG 40's *Erg Staffel* which had been in existence since October 1940. In summer 1942 KG 40 was expanded again, with V *Gruppe* (Ju 88C-6), a *Zerstörer* unit comprising 13. to 16. *Staffeln*.

KG 40 made its operational debut during Operation *Weserübung*, the invasion of Norway and Denmark in April 1940, when Fw 200C-0s of 1./KG 40 (subordinated to X *Fliegerkorps*) were used in the anti-shipping and supply roles from bases at Aalborg and Copenhagen-Kastrup. I/KG 40 was soon brought up to three-*Staffel* status and equipped with Fw 200C-1s, these aircraft subsequently operating against targets in Great Britain from Brest-Guipavas and Bordeaux-Mérignac in France during August 1940 as part of *Generalmajor* Joachim Coeler's IX *Fliegerdivision*. Later, I/KG 40 was subordinated to IX *Fliegerkorps* of *Luftflotte* 2.

With the formation in March 1941 of *Fliegerführer Nord*, *Lofoten*, and *Atlantik*, KG 40 (under the command of *Fliegerführer Atlantik* with Headquarters subordinated to *Luftflotte* 3 at Lorient in France) concentrated on anti-shipping and reconnaissance operations against maritime targets in the Atlantic Ocean. Thus, much of I/KG 40's operations involved the *Gruppe*'s Fw 200Cs acting as *Fuhlüngshalter* (shadow aircraft) on behalf of the *Befehlshaber der U-Boote* (Flag Officer Submarines). These operations continued during

Maintenance work underway on II/KG 40 He 177s at Bordeaux-Mérignac before an operation with Hs 293A-1s. Five of the six aircraft visible are each carrying two of these weapons on their underwing racks.

A Focke-Wulf Fw 200C-1 of KG 40, at Gardermoen in Norway during 1940. Note the weapon rack outboard of the outer engine nacelle.

1942, one notable success being the location of supply convoys *PQ 14* and *PQ 15* as they made their way to Murmansk and Archangel respectively during March 1942. For a while, 2./KG 40 left the domain of *Luftflotte* 3 and was subordinated to *Fliegerführer Nord* at Trondheim, Norway.

In June 1942 the Do 217E-equipped II *Gruppe* was discharged from KG 40 and, after an intensive retraining phase on the He 177 at Lechfeld, became operational as I/FKG 50 (itself formed at Brandenburg-Briest in late 1942 from a cadre of 10./KG 40). Earlier, on 8 September 1942, 1./KG 40 (less its Field Service Company) had arrived at Fassberg for conversion. By 3 October, the technical personnel had completed their instructional course, and 1./KG 40 began to take shape as an He 177 operator. The first five He 177A-0s had arrived at Fassberg on 29 September, followed by the first seven A-1s during October.

Retraining of the aircrews continued until November 1942, but was slightly delayed by the introduction of technical modifications to the He 177s on strength. A two-month course in bombing and aerial gunnery began in December 1942. At this time, 1./KG 40's aircraft were brought up to Equipment Condition 'A' (short-range bomber) and the *Staffel* attached to I/FKG 50 for operations in the Stalingrad area in support of the German 6th Army during January 1943; but the end of 'Fortress Stalingrad' came before the KG 40 crews could be 'blooded'. Ironically, the Fw 200Cs of I/KG 40 had also found themselves on the Eastern Front, grouped together in the specially-formed *KGrzbV* 200 to help supply the German troops. The unit was disbanded in March 1943

Back in the West, KG 40's contribution to *Fliegerführer Atlantik* was reduced to the Fw 200Cs of III/KG 40 and some Ju 88C-6s of 13./KG 40 in occupied France, the latter type operating against RAF Coastal Command anti-submarine aircraft over the Bay of Biscay.

At the end of 1942 elements of I *Gruppe* as well as 8. Staffel began retraining on the He 177, while simultaneously crews of II *Gruppe* began their conversion onto the Do 217E-5 at Soesterberg in The Netherlands. By exchanging 8. *Staffel* with 2. *Staffel*, III/KG 40, until then operating Fw 200Cs alongside He 111H-6s, became a purely Fw 200C-equipped *Gruppe*. The move anticipated a new role for the Fw 200C in addition to its use as a long-range maritime reconnaissance platform: the intensive instruction of Hs 293 guided missile operators.

Due to various technical shortcomings, the He 177s were grounded from 7 February to early May 1943, by which time only four aircraft had been cleared to resume flying. These aircraft were then used for further training of own aircrews and

A parked He 177 A-3 of KG 40.

combat observers until November 1943. At Fassberg, 2./KG 40 had arrived on 1 February 1943 prior to conversion to the He 177. However, due to the grounding order effective from 7 February the first He 177A-1 could only be ferried on 26 May; 11 A-3s followed by 20 October.

By 15 August 1943 the retraining of aircrews was fundamentally complete. As the intended operations would involve the use of *Kehl*-equipped He 177s, the guidance operators had to be instructed by *E-Lehr Kdo* 36, but this training was interrupted after just four weeks because III/KG 40 was to reform at Garz. Consequently, it was only after some delay that training in the art of using the *Kehl* guidance equipment was restarted at Schwäbish-Hall and Giebelstadt, on 1 December 1943. This was followed by practical bombing training at Garz, and thus 2./KG 40 was not considered ready for operations until 15 April 1944. Meanwhile, on 12 December 1943 an order was received stipulating the re-equipment of all available He 177A-3s for operations against Great Britain. Four days later 1./KG 40 transferred to Châteaudun in France.

The Technical School of *Luftflotte* 2 responsible for training ground personnel at Fassberg had two He 177s for instructional purposes, these being the second A-0 built by Arado and an A-1. In June 1943, IV/KG 40 also had only two He 177 training aircraft, both A-0s, to instruct its crews on this new long-range bomber. The number was increased during the second half of 1943 with the arrival of 12 *Kehl*-equipped and several other He 177s; but due to the aforementioned grounding of all He 177s between February and May, training could not restart until October. A good seven months had been lost. Early in 1944, the training unit was transferred to Lechfeld.

A more serious problem was the lack of operationally experienced aircrews for instructional purposes. Due to the high loss rate II/KG 40 could not transfer any experienced crews to IV/KG 40 until March 1944, when two crews were made available for this vital task. As a result of this personnel shortage, 24 aircrews had to be handed over to I and II/KG 40 after only 15 hours of instruction on the He 177. Not only that; none of the new crews could complete their 'special weapon' training while at IV/KG 40 for lack of a proper bombing range.

On 14 April 1944 IV/KG 40 had a total of 35 He 177A-0/-1/-3s, of which only 13 were serviceable. There were six instructors to train the young crews on the Fw 200, and 10 others for the He 177 – a total of just 16 instructors for no less than 80 student crews! Matters were made worse by the low serviceability of the He 177s used for training purposes due to the lack of replacement powerplants, and the loss of new-build He 177s as a result of enemy air raids.

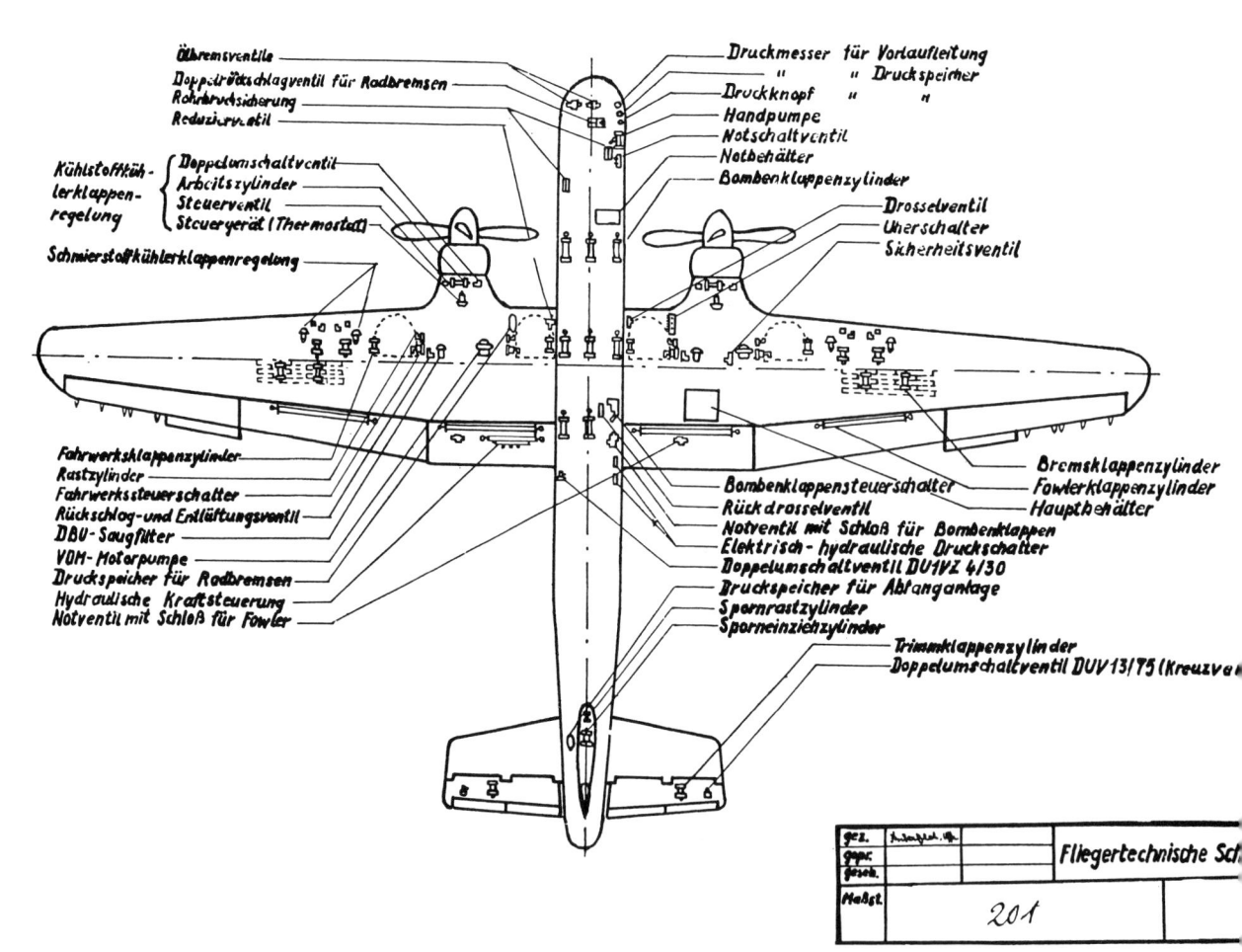

Instructional material used by Fliegertechnische Schule 2. *The drawing shows the hydraulic devices on the He 177.*

A new operational phase for I/KG 40 began shortly after the transfer of 1. *Staffel* from Fassberg to Châteaudun in December 1943. Before that, on 15 October, *Feldwerft-Abt zbV* 1 had been transferred aboard two transport trains to Bordeaux-Mérignac in southern France and had its hands full with He 177s arriving almost daily from Germany. In the meantime, *Feldwerft-Abt zbV* 2 had established itself at Châteaudun and was busy looking after the operational aircraft of I/KG 40, as well as those of I/KG 100 that shared the base.

The Technical Officer of the base, *Oberleutnant* Reper, who had a very positive opinion of the He 177, suggested several interesting improvements to the aircraft. One of these concerned a

Aircrew of 4. and 5./KG 40 minutes before a training flight in their He 177A-5s from a rather wet Bordeaux-Mérignac in south-west France.

KG 40's imposing flight operations building at Bordeaux-Mérignac, with vehicles of the airfield fire service detachment in the foreground.

Düppel* installation that could be operated by the aft dorsal gunner; another was a special curtain that prevented blinding of the aircrew by enemy searchlights. He had also seen to it that several of the aircraft based at Châteaudun and used mainly during daylight hours, were armed with a single 20 mm MG 151/20 cannon in the upper nose gun position.

By mid-January 1944 the most important preparations for participation in the planned Operation *Steinbock* (Capricorn), the so-called

---

* Metal foil strips dropped by bombers to 'blind' enemy radars. Known as *Window* to the RAF; named *Düppel* by the *Luftwaffe* after the town near the Danish border where they were first found.

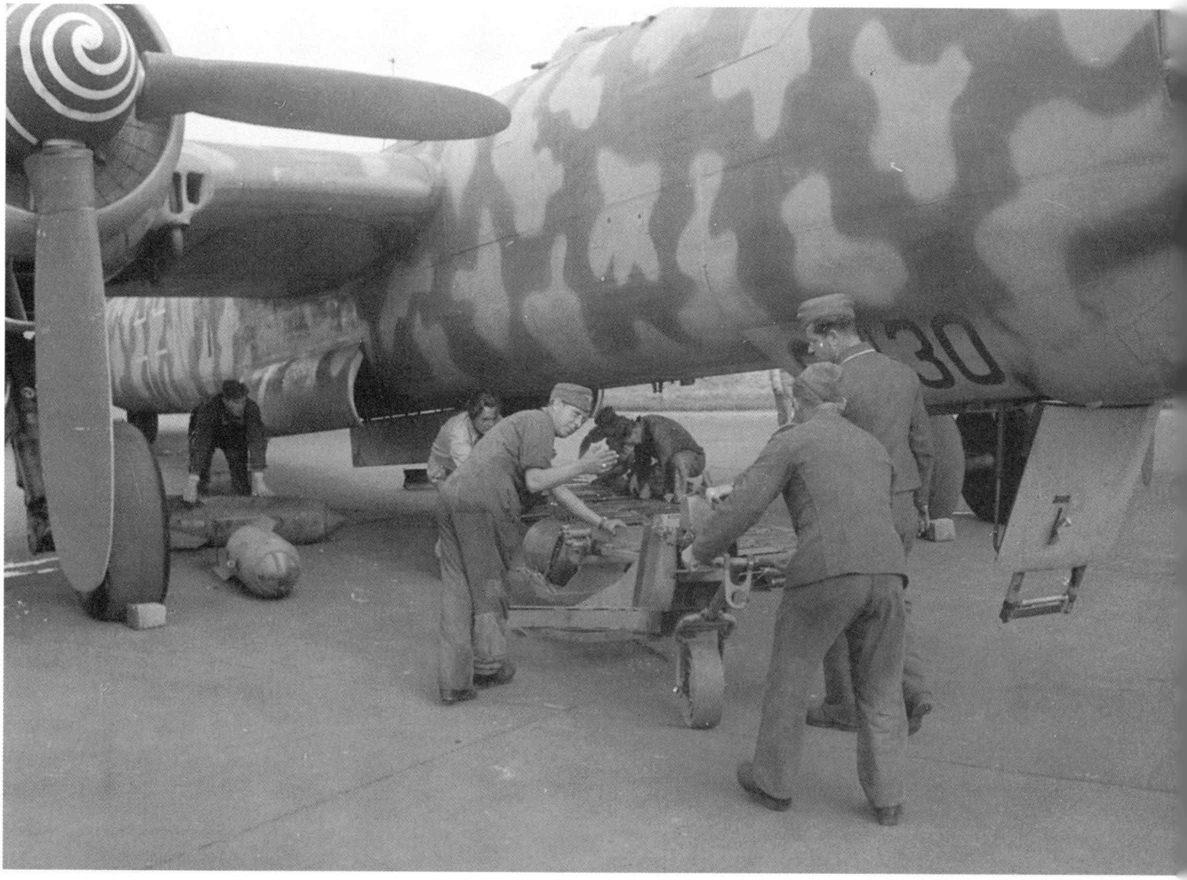
30

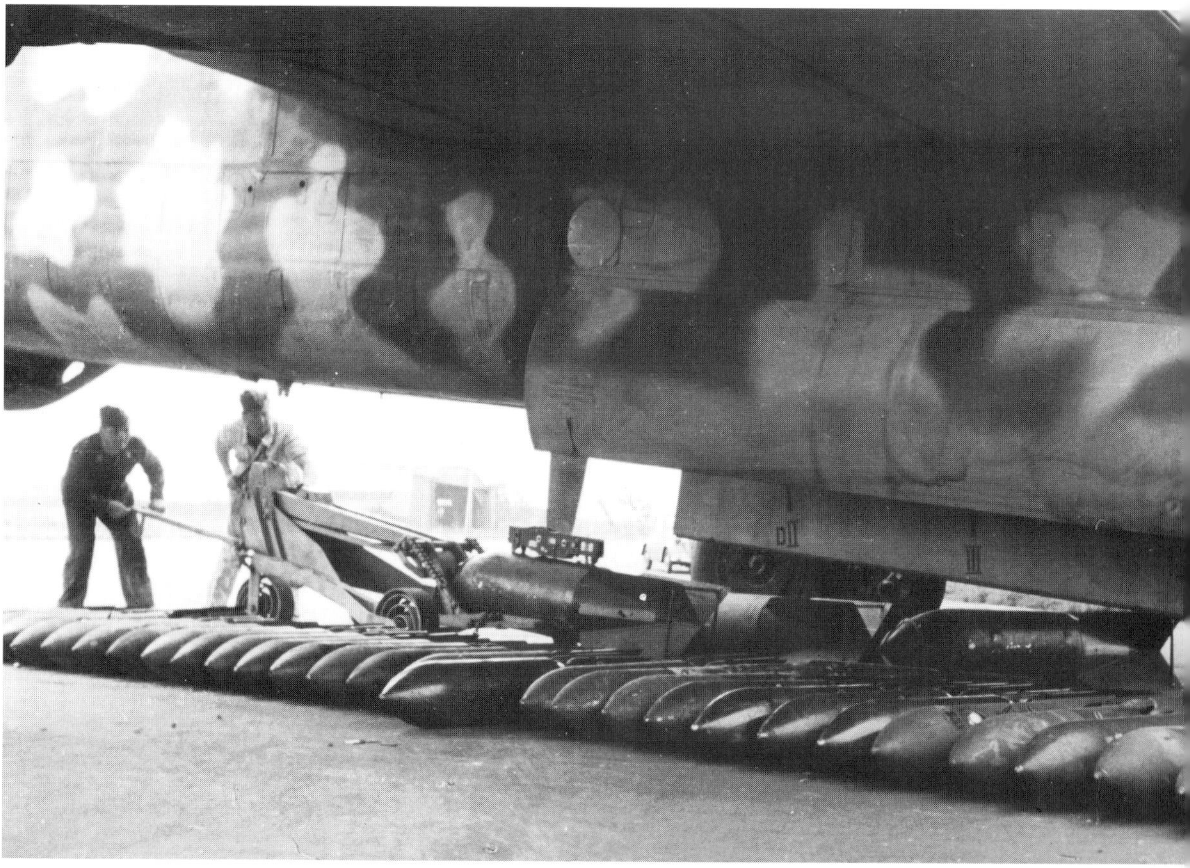
DII

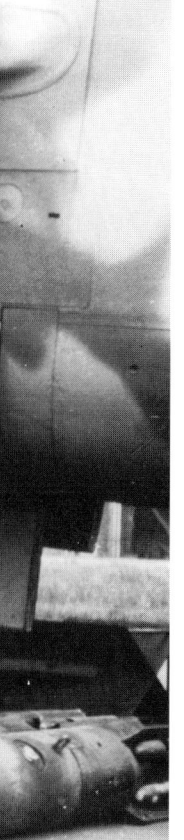

Loading an He 177A-3 of KG 100 with SC 250 and SC 500 bombs.

'Little *Blitz*' over England, had been completed by I/KG 40. The *Gruppe* was also strengthened with the arrival of additional crews from elements of II/KG 40 and I/KG 100, bringing together a total of 32 He 177A-3s under the command of *Hauptmann* Meierhofer.

During the period 21-29 January 1944, three nocturnal bombing raids were carried out on targets in the London area at a cost of 12 He 177A-3s. A further four aircraft were lost due to engine fires over France. Three of the crews managed to bale out and parachute to safety, but one aircraft (Wk-Nr 332221) crashed with the loss of all aboard. On 24 January 1944 another A-3 (5J+HL/Wk-Nr 332217) was lost, this time to German Flak gunners near Antwerp; while another aircraft fell victim to British night-fighters. By 7 February the KG 100 combat group concentrated in northern France (redesignated I/KG 100 and comprising elements of I/KG 40 with 3./KG 100) had just 20 operational He 177s left. After several more nocturnal bombing raids over England, during which a number of I/KG 40 crews flew as an integral part of I/KG 100, both *Gruppen* parted company again.

On 7 February 1944, after handing over all of its operational aircraft, I/KG 40 transferred back to Fassberg and began re-equipping with the new He 177A-5. By 15 April, the entire *Gruppe* had also finally completed its training on the *Kehl* guidance equipment. At that stage, I/KG 40 could field 36 He 177A-3s, of which 21 were *Kehl III*-equipped for use in conjunction with Hs 293s. However, only 25 of the 36 aircrews were fully-trained.

During the first days of June 1944 I/KG 40 transferred from Orléans-Bricy to Toulouse-Blagnac, their home base having been rendered unusable by virtue of several Allied bombing raids in the run-up to the D-Day landings. On 26 June 1. and 2. *Staffeln* could both field combined total of just 12 He 177A-3s at Toulouse-Blagnac. Together with a similar number of aircraft serving with II/KG 40 at Bordeaux-Mérignac and one He 177 attached to *Stab*/KG 100 at Toulouse-Francazais, this gave a total of just 25 He 177A-3s and A-5s in the West. Early in July 1944, after just three operations on the Invasion Front which cost it 10 crews and 11 He 177A-5s, I/KG 40 had to be transferred back to Celle in Germany for rest and replenishment.

II/KG 40 had a completely different evolution.

Last-minute discussions before an operational flight by II/KG 40 crews from Bordeaux-Mérignac.

Having formed the basis of I/FKG 50 in late 1942, on 25 October 1943 I/FKG 50 once again became II/KG 40 and transferred from central Germany to Bordeaux-Mérignac that same month. The new He 177A-3-equipped II/KG 40 replaced the previous, Do-217E-equipped II/KG 40 which had been redesignated V/KG 2 'Holzhammer' in June 1943.

After a thorough check of its Heinkel aircraft, the new II/KG 40 conducted its first operational sortie on 21 November 1943. At about 12:15 hours, a total of 25 He 177A-3s took off for an attack on the 64-ship convoy *SL.139/MKS.230*, making its way from Great Britain to Sierra Leone and then North Africa. After more than four hours in the air, the convoy, protected by four destroyers and several corvettes, was spotted at 16:30 hours in a position 18°W of Brest. Despite the most difficult weather conditions, the II/KG 40 combat group turned towards their quarry. An attack in close formation was out of the question and so the aircraft had to approach their targets individually. The results were dismal; apart from two Hs 293s released by the crew of *Hauptmann* Nuss, none of the other guided weapons found their target!

The two successful Hs 293s scored direct hits on two large freighters of 8,132 and 10,165 tonnes (8,000 and 10,000 tons) respectively. Against that, three He 177A-3s were lost, and a further four were damaged by well-directed anti-aircraft fire from the convoy escorts.

On 26 November, II/KG 40 attacked a convoy in the Mediterranean Sea with no less than 21 aircraft. This operation cost the lives of six crews, including the *Gruppenkommandeur* Major Mons (F8+DX/Wk-Nr 535677) and *Hauptmann* Nuss (F8+BP/Wk-Nr 535684). Two other A-3s crashed on their return flights. The other side of the balance sheet, as reported by the returning crews, listed the sinking of a destroyer and two frigates totalling 28,461 tonnes (28,000 tons). During the ensuing aerial combat the He 177 gunners shot down no less than 10 Allied fighters of which three were accounted for by the crew of *Hauptmann* Eidhoff.

Operation *Shingle*, the landing of the US VI Corps at the Italian ports of Anzio and Nettuno, began on 22 January 1944. The *Luftwaffe* opposition, in addition to the Do 217s of II/KG 100, Ju 88s of KG 1 'Hindenburg' and the aircraft of I and III/KG 26 'Löwen', also included the

Hs 293-carrying He 177s taxy out at Bordeaux-Mérignac at the start of another mission against Allied shipping in the Atlantic Ocean.

He 177A-3s of II/KG 40 armed with PC 1400X glide-bombs. During the first 24-hour period after the Allied landings, II/KG 40 flew a total of seven operations, resulting in the sinking of a 3,050 tonne (3,000 ton) freighter by the aircraft of the new *Gruppenkommandeur*, *Hauptmann* Rieder. In addition, glide-bombs damaged an Allied cruiser and a destroyer for the loss of eight He 177A-3s and five complete crews during the last three operations.

By then, the losses incurred by II *Gruppe* totalled 24 aircraft; an additional six aircraft had been delivered to the field repair workshops at Toulouse-Blagnac. This situation could not continue, and on account of the losses suffered during the first five attacks in January 1944 II/KG 40 was granted a few days rest at the end of the month.

One positive aspect of the operations was that the operational safety and reliability of the He 177A-3 had been improved, doing away with the need for the usual six- and 12fi-hour control checks. The regular 25-, 50- and 75-hour inspections were now completely sufficient, with special attention being paid to servicing of the coupled powerplants after 50 flying hours. According to the technicians, the He 177 serviceability rate of II/KG 40 was frequently in the order of 80 per cent; a great improvement over the 30 per cent or so recorded during the *Gruppe*'s training phase, when flying operations were noticeably affected by moisture in the air which led to frequent accidental earthing of onboard electrical equipment.

In contrast to the situation with I/KG 40, only one aircraft assigned to II/KG 40 was lost due to powerplant failure. During operations against Great Britain there had been numerous powerplant problems, caused mainly by the undertrained aircrews overstraining the engines. On the positive side, the Bordeaux-Mérignac-based *Gruppe* had carried out the first He 177 long-range flight (lasting 12fi hours) and proposed to increase the aircraft's range still further by using 900-ltr (198 Imp gal) underwing auxiliary fuel tanks. Despite this overload, but obviously helped by the even stressing of both powerplants during the long-range flights, it had proved possible to operate engines for up to 115 flying hours without any problems.

An He 177A-5 (KM+UU/Wk-Nr 550071) of II/KG 40 at Nancy, before continuing its flight to Châteaudun.

A newly-delivered He 177A-5 (KM+UD/Wk-Nr 550054), complete with underwing weapon racks, at a KG 40 base. This aircraft was later assigned to 6./KG 40.

One of II/KG 40's He 177A-5s (KM+UJ/Wk-Nr 550060) undergoing servicing, with the forward dorsal turret receiving special attention.

A combined operation involving II/KG 40 He 177A-3s and some III/KG 40 Fw 200C-4s took place in February 1944. This cost II *Gruppe* its leading crew when *Gruppenkommandeur Hauptmann* Rieder's aircraft (Wk-Nr 535695) was intercepted and shot down by Allied fighters over the Atlantic Ocean.

At this time the He 177A-3s in service with II/KG 40 were gradually being replaced by the more powerful A-5 model. The *Gruppe* flew its last operation for the time being on 12 February and then began training new crews for the freshly-allocated He 177A-5s. The plan was to re-equip II/KG 40 with 30 A-5s, and transfer the displaced A-3s to training units. However, deliveries of A-5s were somewhat sporadic: only eight aircraft had arrived by 23 March 1944, the balance of the *Gruppe*'s strength being made up of 30 A-3s. A minor problem then arose because the prescribed offensive load of the *Kehl IV*-equipped A-5s meant that the aircraft did not have the necessary operational range; all of these aircraft had to be retrofitted to *Kehl III* standard to resolve the problem.

On 16 March the *Fliegerführer Atlantik* command post, until then under the command of *Generalleutnant* Kessler, was replaced by X *Fliegerkorps* under *Generalleutnant* Holle. The new Commanding General inspected II/KG 40 as well as *Feldwerft-Abt zbV* 1 on 21 March. It was quite a day; just as one He 177A-5 (KM+UO/Wk-Nr 550065) was taxying for take-off, it was strafed by seven raiding Allied fighters and burst into flames.

A different kind of excitement occurred three days later, when an He 177A-3 (Wk-Nr 535692) veered off the runway at Bordeaux-Mérignac while taxying. The ground technician at the controls tried to keep the big aircraft level by alternatively increasing the throttle on both powerplants; as a result, the aircraft gained ever more speed until it started to skid. The hectic ride ended abruptly when an undercarriage leg broke off.

Only three days later, on 27 March, the Bordeaux-Mérignac base was attacked by 160 USAAF bombers, which destroyed some hangars and damaged the runway in several places. At that time, II/KG 40 possessed 27 He 177A-5s. But as the air raid warning had mentioned 600 heavy bombers escorted by 250 fighters, no attempt was made to carry out an emergency take-off to clear

At rest between missions, an He 177A-5 (6N+EM/Wk-Nr 550023) of 4./KG 40 . . .

. . . The same aircraft with engines fired up, taxying out for another mission. Note the extensive flap area on each wing.

the base. Amongst the aircraft hit while on the ground, three A-5s were so badly damaged that they had to be broken up for spares, but another five A-5s were repaired within a matter of weeks. The next day, the operational base of I/KG 100 was again attacked by Allied bombers, 128 in number, while low-level attacks by fighters resulted in further losses.

The following month was almost peaceful, and II/KG 40 transferred to an airfield near Gotenhafen for training purposes. It was planned to carry out an exercise in co-operation with the *Kriegsmarine*. In early May, while on its way to the base, an He 177A-5 (Wk-Nr 550142) ran into bad weather and was lost with the entire crew. Icing was blamed, but it seems an RAF long-range fighter had intercepted and downed the big bomber. Early in June 1944, having suffered a further five losses, II/KG 40 transferred back to Bordeaux-Mérignac, and operations against the Allied D-Day invasion forces off the coast of Normandy began a few days later. It was to prove a very costly undertaking: during the very first attack II/KG 40 lost 13 of its 26 He 177A-5s.

A jacked-up He 177A-5 of II/KG 40 during maintenance work on its port-side main undercarriage unit. Note the two open undercarriage bay doors, one bay for each sideways-retracting oleo.

A formation flypast by eight He 177A-5s.

The attrition continued next day when all remaining II/KG 40 He 177A-5s took off for another attack. Seven were shot down by Allied fighters, and only six – some badly shot up, but kept flying thanks to the skill of their crews – managed to return to base. Other elements of II/KG 40 were also transferred from Lechfeld to Bordeaux-Mérignac early in June 1944 and flew several nocturnal operations from that base. Even the cloak of darkness afforded no protection against superior odds, and several more He 177s were lost; only six aircraft managed to return safely to base.

In just six operations on the D-Day Invasion Front II/KG 40 had lost no less than 14 crews and 20 He 177A-5s – a high price to pay for sinking 68,103 tonnes (67,000 tons) and damaging 32,527 tonnes (32,000 tons) of enemy merchant shipping; in addition to which the crews of II/KG 40 could claim the sinking of 5,794 tonnes (5,700 tons) and the damaging of 28,461 tonnes (28,000 tons) of enemy warships. The balance was even less positive in the air: in fierce aerial combats over northern France the gunners of II/KG 40 managed to shoot down just three enemy fighters.

In August 1944, II/KG 40 and II and III/KG 100 were withdrawn from operations for a period of rest. II/KG 40 was transferred back to Lechfeld and cocooned its He 177A-5s in nearby woods. By the end of the month, a total of 37 *Kehl*-equipped anti-shipping He 177s and a further 155 He 177 bombers in *Luftwaffe* service had been parked at various locations 'until further orders'.

That same month, the He 177 *Kehl-Gruppe* transferred from Bordeaux-Mérignac to Oslo-Gardermoen in Norway. On 15 September all 28 He 177A-5s assigned were based there, parked in full operational readiness in widely-separated 'splinter boxes'; but there was no more fuel for any further flying.

As there were not enough ferry crews at Bordeaux-Mérignac, one He 177A-3 (Wk-Nr 535675) and seven of the newer A-5s had to be blown up as Allied forces advanced through northern France and forced the German forces into retreat. The same fate befell three other *Kehl*-equipped A-5s that happened to be under repair in the field workshops at Toulouse-Blagnac at that time.

There was to be no improvement in the fuel situation during October 1944. Indeed, the shortage was so acute that even essential workshop test flights could not be conducted.

With effect from 6 November 1944, KG 40 began to reform as a *Schnellkampfgeschwader* (Fast Bomber Group), with the exception of 8. *Staffel* which was still equipped with Fw 200Cs and which became Transport *Staffel* Condor. The

Close-up of the nose of a II/KG 40 aircraft fitted with an array of FuG 200 Hohentwiel *ASV radar aerials and an underfuselage weapon rack immediately aft of the ventral gun position.*

He 177 training and replacement element, IV/KG 40, was disbanded, except for 4.(*Kehl*)/Erg *Staffel*. The Ju 88C-6-equipped V/KG 40 had been discharged to form I/*Zerstörergeschwader* (Destroyer/Heavy Fighter Group) 1 as far back as October 1943. Now, all of KG 40's surviving He 177s were neatly parked around the perimeters of various airfields, awaiting their fate.

The He 177 in Service with KG 100 'Wiking'

The *Flugfunker Schul-und Versuchskommando* (Flight Radio Operator Training and Experimental Detachment) was activated in August 1938 and at the end of that month was redesignated *Luftnachrichtenabteilung* (Aviation Signals Detachment) 100. Equipped with Do 17s, Ju 52s and He 111s, this formed the nucleus of *Kampfgruppe* (KGr: (independent) Bomber Group) 100, activated on 18 November 1939 and eventually comprising three *Staffeln* equipped with He 111Hs. In November 1941, after participating in the campaigns against Denmark and Norway, the Low Countries and France, Great Britain and fighting on the Central sector of the Eastern Front, KGr 100 was transferred back to Hannover-Langenhagen. That same month a *Geschwaderstab* (Group Staff/Headquarters) was formed and, on 18 December 1941, KGr 100 was officially reformed as I/KG 100 'Wiking'.

Early in 1942, 2./KG 100 transferred out to become *Erprobungskommando* (EKdo: Proving/Test Detachment) 100; but the loss of the *Staffeln* was more than made up for by the redesignation of III/KG 26 'Löwen' as II/KG 100 and the formation of the all-new IV/KG 100 during the winter of 1941-42; then the formation of III/KG 100 from *Aufklärungsgruppe* (*See*) 126 (Naval Aviation Reconnaissance Wing) in September 1942, although this *Gruppe* reverted to its previous designation just four months later. III/KG 100 was subsequently re-established from parts of *KGrzbV* 21 on 20 April 1943. Some six months later, on 10 October 1943, I/KG 100 was redesignated I/KG 4 'General Wever', and simultaneously the existing I/KG 4 became the new I/KG 100 – only for it to be redesignated as III/KG 1 'Hindenburg' on 31 May 1944.

The story of the He 177's service with KG 4 'General Wever' began with a teleprinter message

This was the fate of most He 177s from late summer 1944 onwards: parked, camouflaged, and waiting to be reduced to scrap.

from the General der *Kampfflieger* (GdK: General of Bombers) in mid-December 1942. I/KG 4, at that time based at Sestchinskaya in Russia and equipped with He 111H-6s, was to be withdrawn from operations on the Eastern Front and transferred to East Prussia by early 1943. Accordingly, the first transport left the base on 30 December 1942 and arrived at Gutenfeld near Königsberg in mid-January 1943. After consolidating at their new base, the entire personnel of the *Gruppe* (except for a small holding detachment) was sent on leave.

On 3 February 1943 an advance detachment was sent to Lager-Lechfeld, followed a short while later by the whole of I/KG 4. The intention was to start retraining the aircrews and ground personnel on the He 177 as quickly as possible, and on 20 February the first members of the *Gruppe* received their transfer orders to *Fliegewaffentechnische Schule* (Aerial Armament Technical School) 2 at Meseburg, while others were posted to Fassberg. Retraining of the aircrews began at the same time, with special emphasis paid to the familiarisation of the pilots and navigators with the *Lotfe* 7D telescopic bombsight, and the radio operators and gunners with the MG 131 machine-gun and the remote-controlled weapon installations. Gunnery training began a little later at Vaerlose in Denmark.

The first He 177s arrived on 18 May 1943; a few pre-production A-0s (still quite suitable for training purposes), followed soon after by an initial batch of A-1s. The first instructor crews from I/FKG 50 reported to Lager-Lechfeld at the end of the month and began to devote themselves to an intensive retraining of I/KG 4 aircrews, who, a short while later, had their first opportunity to fly the He 177A-3.

After numerous local and cross-country flights and intensive instruction of the armourers and technicians on what was a lavishly-equipped aircraft compared to the He 111H-6s they had worked on before, the aircrews began to conduct their first target approach, low-level and single-engined flights. By then, I/KG 4 had a new *Gruppenkommandeur*, *Hauptmann* von Kalckreuth.

It was at this stage that several tragic accidents occurred. During a low-level flight over Lake Ammer, *Leutnant* Moor's aircraft suddenly touched the water's surface and crashed. A few days later, on 31 August 1943, it was the turn of *Hauptmann* Speckart and his crew; their aircraft crashed and disintegrated on the ground. The following weeks were again filled with many training flights over southern Germany.

On 20 October 1943 the aircrews on parade were surprised by the contents of a teleprinter message from the GdK, read out to them by *Hauptmann* von Kalckreuth:

'With effect from 6 October 1943, I/KG 4 'General Wever' has been renamed I/KG 100.'

A side-view of a 2./KG 4 aircraft at Rennes on 20 April 1944.

Early in 1944, after about 10 months of retraining, I/KG 100 began operations with the He 177 as part of the bomber force assembled for Operation *Steinbock*. At about 19:30 hours on 21 January, the big Heinkel bombers of 3./KG 100 (along with aircraft from other units) took off and set course towards London. The 227 bombers of the first wave were followed by 220 bombers in the second wave, attacking early the following morning. The raids cost IX *Fliegerkorps* a combined total of 43 aircraft, of which 25 were shot down by RAF night-fighters and well-directed anti-aircraft fire. The third raid (285 aircraft) caused great destruction in the Surrey Docks area of east London.

It was on these raids that the costly airframe strengthening applied to the He 177 paid dividends. Crews of 3./KG 100 would go into inclined flight immediately after releasing their bombs, gaining speed to reach their home base at Rheine as quickly as possible. Even then, though, the RAF night-fighters managed to intercept and damage two aircraft.

For the bombing raids on England on 21, 22 and 30 January 1944, IX *Fliegerkorps* assembled a total of 732 aircraft. Of these, the surprisingly high number of 101 crews – 14.5 per cent of the available force – broke off their attacks before reaching the target area. Of this number, 74 aircraft (73 per cent) returned to base because of technical defects: 38 pilots reported powerplant problems, 31 cited the failure of various instruments, and five aircraft suffered some kind of proven airframe breakdown or disorder. The remaining 27 crews abandoned the mission prematurely for various other reasons. In many cases, the crews involved were relatively young and inexperienced. The largest number of returnees came from the KG 100 combat group, I/KG 100: no less than 14 of its crews abandoned their mission and returned home early.

More than anything else, most of the pilots flying He 177s initially had no idea about the bomber's prescribed engine revs and highest permissible climbing speeds. The inevitable resulting powerplant overstressing led to no less than seven crashes and engine fires. Other crews undercut the minimum permissible speed, stalled and crashed. Prior to that, problems had arisen due to the sudden move to a new base at short notice, which had left too little time for comprehensive servicing of the A-3s assigned. The

The crew of Leutnant Goetze *in front of their He 177A-3 (5J+KK/Wk-Nr 332206). Although the '5J' alpha-numeric code indicates this to be a I/KG 4 aircraft, it was photographed after being reassigned to I/KG 100 . . .*

. . . For a short while after I/KG 4 ('5J') and I/KG 100 ('6N') swapped Geschwaders *and adopted each other's identity in October 1943, each* Gruppe *used its previously-allocated alpha-numeric code.*

unfavourable airfield conditions and insufficient ground organisation only helped to make the whole situation that much more difficult.

The fourth large-scale raid on London, by 240 bombers, took place on the night of 4 February 1944. This time, only six crews of 3./KG 100 at Rheine took part. The attack was well-planned and executed, with target-marking performed by I/KG 66 pathfinders and screening of the route by the release of bundles of *Düppel*. The He 177A-3s reached London and, after releasing their bomb-loads, the remaining five aircraft went into

THE HE 177 IN MARITIME WARFARE 141

Loading bombs aboard a KG 100 aircraft detailed to take part in an Operation **Steinbock** *raid against targets in Great Britain in spring 1944.*

One of the numerous low-level strafing attacks by USAAF fighters on He 177A-3 and A-5 bombers parked on airfields in southern Germany.

inclined flight at 560 km/h (348 mph) and headed for Münster.

The London area was the target for another *Luftwaffe* raid (200 bombers) on the night of 21 February 1944. The following night, 185 bombers attacked selected targets in southern England.

Two heavy air raids involving a total of 352 *Luftwaffe* bombers took place in early March 1944. On 14 March the target was once again London, the attacking force of almost 190 bombers including He 177A-3s from 1. and 2./KG 100. A few days later, on 18 and 19 March, strong Allied bomber formations attacked the base of 1. *Staffel* as well as the Lechfeld area. These raids caused heavy losses among *Luftwaffe* personnel, as well as the destruction of four He 177s and damage to six others.

A total of 33 He 177s participated in the five major raids on targets in Great Britain during the period 19 March to 19 April 1944, but the number of available aircraft fell dramatically as a result of operational losses and crashes. On 25 and 26 March the He 177 bases were attacked by Allied low-level fighter-bombers, resulting in the deaths of a number of ground personnel and the destruction of three aircraft belonging to I/KG 100. Two days later, Châteaudun was the target of an Allied bombing raid which destroyed or damaged another three He 177s.

The transfer of six He 177s from Châteaudun to Rheine was then ordered at short notice, and this time it was the *Luftwaffe*'s own flak that opened up on the aircraft and damaged three of them, after Münster had reported an incoming attack by six USAAF B-17 Flying Fortress bombers. The flak battery kept on firing even after the He 177s had come down to an altitude of just 200 m (656 ft) and fired more than 100 recognition flares. Incredible as it might sound, the persistent flak gunners did not stop firing at the He 177s even after the aircraft had landed and were taxying in! As a result of the flak battery's attentions, two of the six aircraft were so badly damaged they had to be written-off as total losses.

On 14 April 1944 I/KG 100 could field just 17 He 177s at Châteaudun; a further eight aircraft were under repair in outside workshops. However, of the 17 available only eight were fully serviceable and ready for operations; five had to remain grounded for lack of spares, and another four were simply classified as 'not serviceable'.

Expressed in cold statistics, this gave a technical serviceability rate of 66 per cent, ignoring aircraft damaged by enemy action, otherwise just

The crew captained by Feldwebel Hüppmeier at Minden on 23 April 1944, shortly before their transfer to Châteaudun.

The He 177 in Maritime Warfare   143

Although not visible in this shot, the nose of this 2./KG 100 He 177 bore the name Susy.

A flypast by 2./KG 100's Marga *during Operation* Steinbock *in spring 1944.*

32 per cent! The number of operationally suitable (i.e. fully-trained) personnel was also wholly unsatisfactory; a fact pointed out by Heinkel some time before, but considered to be of little significance by OKL in Berlin. By mid-April, I/KG 100 had just six complete operational He 177 aircrews.

On 19 April the *Gruppe* went into action over London with just six aircraft, of which one A-3 (6N+AK/Wk-Nr 332379) piloted by *Feldwebel* Reiss failed to return. There was to be no respite. On the night of 20–21 April, a force of 130 He 177A-3s again raided the centre of Hull, followed by another raid on Bristol on the night of 26 April.

I/KG 100 flew its last operation in the West with a fourth raid on Portsmouth as part of a force of 101 bombers. The main objective, larger Royal Navy vessels, were to be attacked with PC 1400X

This He 177A-3 of I/KG 100 was photographed over Lake Pleskau in Russia while on a long-range training flight in autumn 1944.

glide-bombs guided by Do 217K-2s of II/KG 100, while a simultaneous bombing raid was intended to disrupt and disperse British anti-aircraft defences.

Early in May 1944, both *Staffeln* of I/KG 100 were withdrawn from operations and readied for transfer from Châteaudun to Germany via Rheine, arriving at Fassberg on 20 and 21 May for a period of rest and replenishment. During this time the *Gruppe* lost at least another three of its operational aircraft due to enemy bombing raids; while another aircraft fell victim to an Allied low-level attack on 9 May.

Until May 1944, I/KG 100 trained its own aircrews. On 1 May this training unit was supposed to be reformed as an operational *Staffel*, but only four serviceable He 177A-3s were on hand for the 10 aircrews. At the end of the month the remainder of I/KG 100 at Fassberg and Lechfeld formed the nucleus of III/KG 1 'Hindenburg'. It was at this time that both I and II/KG 100 finally received their full allocation of He 177s.

Parked on the Allied-occupied Châteaudun airfield in September 1944, this He 177 of KG 100 bears the name Edith *on its forward fuselage.*

Two views of an He 177A-5, probably a KG 100 aircraft, left behind at Châteaudun as the Allied forces advanced through France in the second half of 1944.

To begin with, in 1943 II/KG 100 flew He 111 and Do 217 bombers from its base at Garz on Usedom Island off Germany's Baltic coast; its retraining on the He 177 did not start until December 1943. In February 1944, 4./KG 100 arrived at Aalborg in Denmark, followed a month later by the *Gruppenstab* and 5./KG 100. The *Luftwaffenführungsstab* expected 4. *Staffel* to be ready for operations by 1 May 1944, and 5. *Staffel* by 1 July, but the first He 177A-3s for 4. *Staffel* only arrived in January of that year. One month later, the A-3s were suddenly and unexpectedly exchanged for A-5s. As for 5. *Staffel*, it received the A-5 model from the outset, beginning March 1944.

By mid-April there were 31 *Kehl IV*-equipped He 177A-5s at Aalborg, of which only 11 were ready for operations. Various technical faults in powerplants and the incomplete *Kehl IV* installations on all 31 He 177A-5s resulted in new delays.

On 5 May *Hauptmann* Bodo Mayerhofer was appointed as the new *Gruppenkommandeur* IV/KG 100, the training and replacement element of KG 100 based at Giebelstadt which had begun its retraining on the He 177 in October 1943. In late January 1944, 10. *Staffel* received its re-equipment

The forward fuselage of an He 177A-5 (6N+HM/Wk-Nr 550043) of 4./KG 100. The lack of an underfuselage weapon rack greatly enhanced the field of fire of the ventral MG 131 machine-gun.

THE HE 177 IN MARITIME WARFARE   147

Reconditioned and repainted, this He 177A-5 wears the '6N' code of KG 100 and was photographed during a workshop test flight over Denmark.

orders and on 31 March listed 15 instructor crews for a total of 58 student crews. Of the latter, 35 were retrained on the He 177. In addition to Giebelstadt, IV/KG 100 also trained at Frankfurt am Main and Schwäbisch-Hall.

During mid-1944, 4. and 5. *Staffeln* continued their training at Aalborg, while the instruction of 8./KG 100 on the He 177 went on at Fassberg. The training course at Fassberg had started on 20 September 1943 and should have been completed by 1 May 1944. Prior to their retraining, 8./KG 100 had flown Do 217s from Istres near Marseille, before being transferred to Schwäbisch-Hall.

Despite the complete lack of any technical support, within a few days the *Staffel* received an allocation of one He 177A-0, seven A-1s and 13 He 177A-3s. Apart from the A-3s the other aircraft were completely worn-out, but all the equipment and expertise necessary for the now essential checks was missing. Things were no better at Burg bei Magdeburg where, due to a lack of hangar space, 10 A-3s were parked outside and exposed to the damp and cold autumn climate.

Even in December 1943, part of the required technical support was based near Burg bei Magdeburg and the remainder at Schwäbisch-Hall, while in the meantime 8. *Staffel* had been transferred to Fassberg. It took two months before the unit could acquire some vehicles; the vitally important special tools for servicing work arrived

A trainee aircrew receive final instructions prior to a flight aboard a 5./KG 100 He 177A-5 (6N+HN/Wk-Nr 550136).

The view from a hangar roof at Fassberg. The mottled grey camouflage on the He 177's upper surfaces and night-black tail, fuselage sides and undersurfaces proved highly effective during nocturnal operations.

The tail gunner, Gefreiter Dops, poses by his 'office' in an He 177A-5.

five months after the unit first received its re-equipment orders.

During the winter months, Fassberg was closed for operations due to flooding. When the weather cleared again, there was a daily average of six hours of enemy air raid alarms. Other than that, all training activity was a long drawn-out affair; the aircraft were spread out over an area eight kilometres (five miles) from end to end and due to the shortage of fuel for the prime movers, a lack of starting devices and the bad ground conditions, they were often moved only with great difficulty to their take-off positions.

The training schedule consisted of instrument-

Powerplant maintenance on an He 177A-5 (6N+KM/Wk-Nr 550045) of 4./KG 100.

and target-approach flights, with special emphasis on low-level attacks, but also on single-engine flying. Inevitably there were losses. On 9 May 1944 an He 177A-5 (6N+DM/Wk-Nr 550006) made a forced landing due to engine failure. Only five days later, the A-5 piloted by *Hauptmann* Mayerhofer (6N+BC/Wk-Nr 550141), the recently appointed *Gruppenkommandeur* IV/KG 100, crashed immediately after take-off, killing the entire crew.

During May, 6. and 8. *Staffeln* were exchanged against each other, resulting in II/KG 100 being equipped entirely with He 177A-3s and III/KG 100 with He 177A-5s. By the end of the month II/KG 100 had almost completed its training, although the first operation, a large-scale raid on shipping near the southern coast of England on 30 May 1944, had to be abandoned; the enemy had received reliable information about this planned attack.

By then, the growing fuel shortage was taking its toll on the training programme, while the numerous aircraft crashes inevitably reduced unit strength. The following causes were identified:

- Defects in generator drive (46.6 per cent)
- Powerplant faults (33.5 per cent)
- Pilot error (13.3 per cent)
- Defective automatic flight controls (6.6 per cent)

In mid-September 1944 II/KG 100, based at Aalborg under its new *Gruppenkommandeur Hauptmann* Molly, had a total of 44 *Kehl*-equipped He 177A-5s – but there was no more fuel for flying. III/KG 100 was already in the process of disbanding when, on orders from the *Luftwaffenführungsstab*, II *Gruppe* was also instructed to disband by 11 October, except for one *Staffel*. A few days before that, 4./KG 100 at Aalborg had 13 He 177A-5s on strength (one of which was temporarily at Garz), while 5./KG 100

A formation flight of eight He 177A-3s of II/KG 100 over Aalborg in Denmark, circa September 1944.

One of the many parked and camouflaged He 177A-5s at Aalborg in summer 1944.

listed 12 serviceable aircraft. A further two A-5s (6N+AC/Wk-Nr 550133 and 6N+CC/Wk-Nr 550127) belonged to the *Gruppenstab*. As for 6./KG 100 at Fassberg, it possessed 15 He 177A-5s, of which all bar three (undergoing re-equipment work) were cleared for operations. But, things being what they were, II/KG 100 with its serviceability rate of 90 per cent managed to clock up a mere four hours of flying time during the whole of September 1944!

The He 177s were kept in full operational order throughout November and December 1944. In fact, although any possibility of their being used in an offensive capacity had virtually disappeared, II/KG 100 was saved from disbandment until 2 February 1945. Only from that day onwards were the He 177A-5s, so carefully maintained in operational order, suddenly discarded. The big bombers were cannibalised, their powerplants removed, and the airframes gradually destroyed, while the *Gruppe* personnel were made available to *Luftwaffe* administration units for ground combat duties. The same sad fate also befell IV/KG 100.

The He 177 in Service with *Wekusta/ObdL*

A study prepared by the *General der Aufklärung* (GdA: General of Reconnaissance) dated 16 February 1943, which envisaged the future allocation of *Luftwaffe* long-range reconnaissance aircraft, listed the Ar 234 and He 177 alongside the Ju 88D, Ju 88T, BMW 801TJ-powered Ju 188 and the Do 217 for day/night reconnaissance tasks.

According to this study, after the completion of 256 Ju 88Ds during the first part of 1943 production was to switch to the Ju 88T-1 beginning in March. In November 1943 the T-1 model would be joined by the more powerful Ju 188, 33 of which would be re-equipped for reconnaissance tasks by 31 December 1944.

The largest reconnaissance element next to the 306 Ju 88s was to comprise 517 Do 217s – at least during the planning stage. Then came 71 Ar 234s and 69 He 177s for 1943, followed by another 86 aircraft up to the end of 1944. Production of the He 177 optimised for long-range reconnaissance was to start with three aircraft in March 1943, and increase to a maximum of 10 aircraft per month. For 1944, the GdA staff planned for the delivery of seven or eight He 177s per month, each aircraft being fitted with a comprehensive automatic aerial camera installation.

This plan was changed on 22 June 1943, when it was decided not to build the He 177 equipped with automatic aerial cameras in series but to produce that installation as a Standard Equipment Set instead. Experience had shown that offensive reconnaissance took second place to weather reconnaissance.

The main *Wetterkungdungstaffel* (*Wekusta*:

The two gunners and the radio-operator pose by their He 177A-5 (6N+DN/Wk-Nr 550131) of 5./KG 100 at Aalborg in Denmark in summer 1944.

The crew of Feldwebel Niederstadt *(including the aforementioned two gunners and radio-operator) in front of 6N+DN shortly before the final parking of KG 100 He 177A-5s due to the lack of aviation fuel.*

THE HE 177 IN MARITIME WARFARE   153

Leutnant John and his crew (II/KG 100) pose in front of the same He 177A-5.

Another view of 6N+DN 'out to grass' at Aalborg in summer 1944.

An He 177A-5 abandoned in the woods on the outskirts of Fassberg airfield in 1945.

Weather Reconnaissance Squadron) eventually equipped with the He 177 was formed in May 1941 from parts of *Wekusta 1/ObdL* (itself originally formed from *Wekusta/ObdL*). Initially, *Wekusta 1/ObdL* flew He 111s and Ju 52/3ms, then Ju 88s. Later, the unit was deployed to Nantes and then Brest, from where its aircraft flew weather/long-range reconnaissance sorties up to the Irish Sea and over the Atlantic Ocean, to about the latitude of Portugal.

From July 1943 onwards, *Wekusta 1/ObdL* took over the operational area hitherto covered by *Wekusta* 51 (formed in summer 1939 to provide weather reconnaissance for *Luftflotte* 3). In April 1944, *Wekusta 2/ObdL* was withdrawn from operations and re-equipped with the He 177. Operations over the Atlantic Ocean with the new aircraft began in June 1944.

Flying over the North Sea, this He 177 caught sight of a surfaced U-boat.

THE HE 177 IN MARITIME WARFARE   155

This He 177A-5 was equipped with the FuG 200 Hohentwiel ASV radar, hence the nose array of aerials. The rear-view shows clearly the sideways-hinged canopy to enable access to and exit from the tail gunner's compartment.

In summer 1944, 1. and 2. *Staffeln* of *Fernaufklärungsgruppe* (FAGr: Long-Range Reconnaissance Wing) 5, hitherto equipped with Ju 290As, received their disbandment orders. Later, both *Staffeln* were to be re-equipped with the He 177 or possibly the Me 264, while the bulk of the reconnaissance formations were to receive the Ju 188. A few months later, on 7 September 1944, *Wekusta* 2/*ObdL* was disbanded on orders from the OKL. Before that, the unit had been transferred from Bordeaux-Mérignac to Burg bei Magdeburg. At that time, *Wekusta* 2/*ObdL* had a total of 17 He 177s of which two A-3s (Wk-Nr 332628 and Wk-Nr 332629) were even armed with twin-barrel tail defensive armament.

Following disbandment, a special engine column arrived at Burg bei Magdeburg from Daimler-Benz to remove and professionally conserve the powerplants from *Wekusta* 2/*ObdL*'s He 177s. The engineless aircraft were then dismantled.

It is not known if the planned reformation with Ju 188 reconnaissance aircraft ever took place.

An He 177A-3 armed with an MG 131Z in the tail gun position.

Chapter 4

The Long-Range Bomber Without a Chance

The RLM Demands Four Separate Engines

Further successful development of the He 177 was dependent on solving the powerplant problems that had dogged the programme from the very beginning. With this in mind, and in case the Daimler-Benz DB 610 coupled powerplant proved unable to reach its full operational status soon enough, in 1943 the Heinkel works proposed to the RLM a project to fit one pattern aircraft with four individual radial engines. Several other interim solutions featuring different wing area/powerplant combinations were also put forward for consideration:

- 108.00 m≈ (1,162.50 ft≈) wing area and four BMW 801E radials
- 108.00 m≈ (1,162.50 ft≈) wing area and four DB 603E inlines
- 133.00 m≈ (1,431.60 ft≈) wing area and four BMW 801E radials
- 133.00 m≈ (1,431.60 ft≈) wing area and two DB 613 coupled inlines

By using the He 177A-7 airframe, then in the planning stages, the following configurations and performance estimates for a future He 177 bomber capable of relatively high speeds were thought possible:

- Series-size wing area and two DB 610s: 490 km/h (304 mph)
- Smaller wing area and four BMW 801Es: 540 km/h (336 mph)
- Larger wing area and four BMW 801Es: 525 km/h (326 mph)
- Larger wing area and four DB 603Gs: 540 km/h (336 mph)

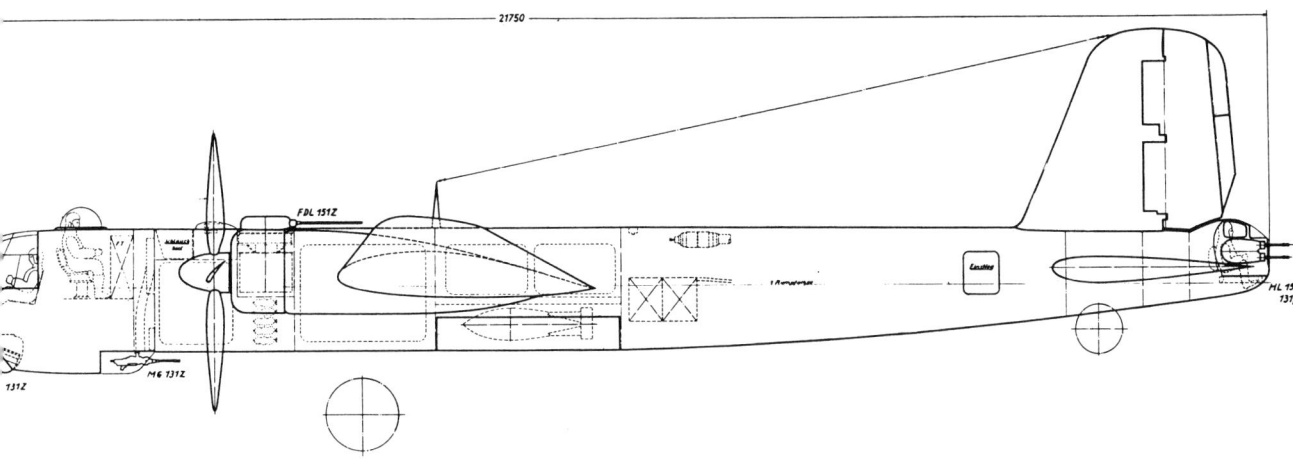

Above and overleaf:
Side- and three-view drawings of the He 177A-7 with DB 610 powerplants, dated 18 August 1943.

	# Typenblatt	**He 177** A-7

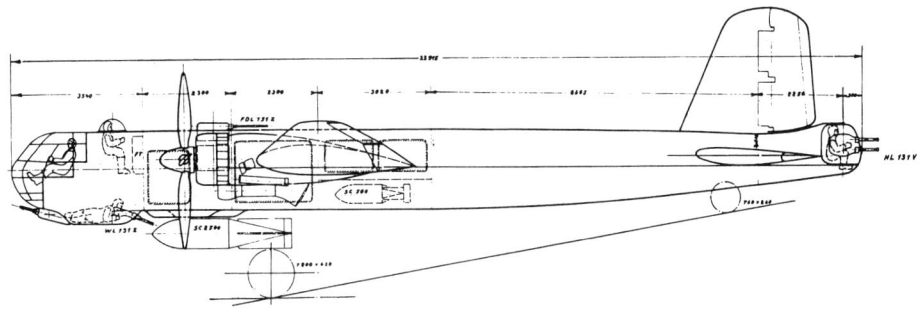

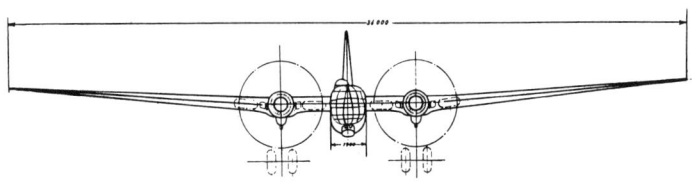

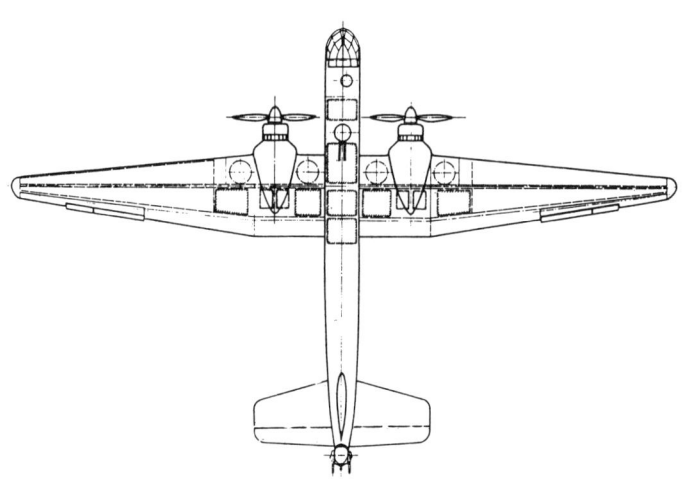

| M: 150, M: 300 | 2 × DB 610 | | WIEN-SCHWECHAT
den 18.8.1943 |

The Long-Range Bomber Without a Chance 159

He 177A-5 GP+RY/Wk-Nr 550256 after capture by the French Resistance.

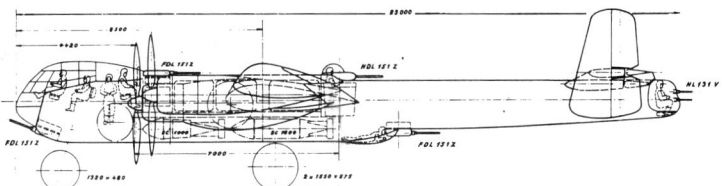

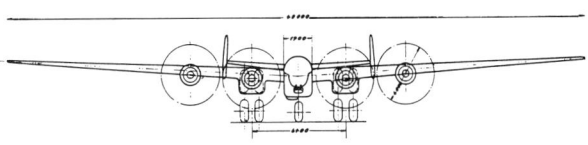

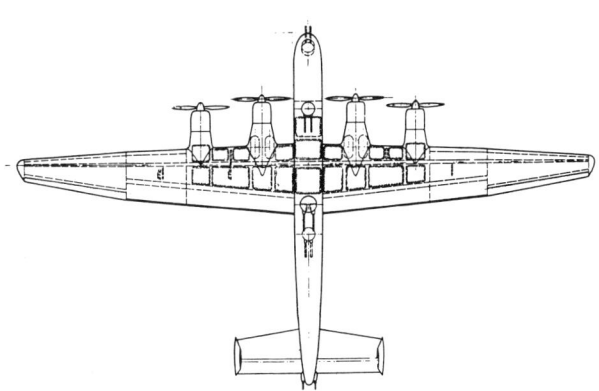

The He 177B-5 (Stage I) long-range bomber.

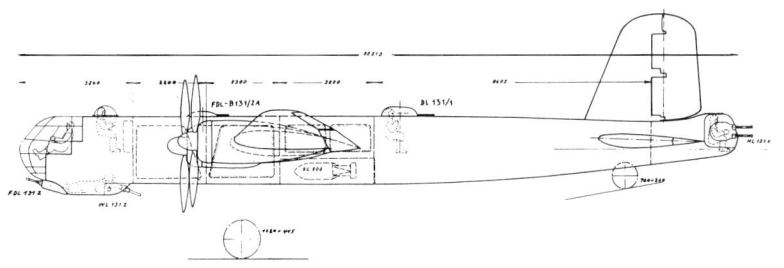
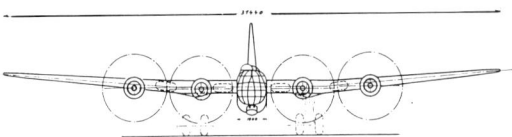
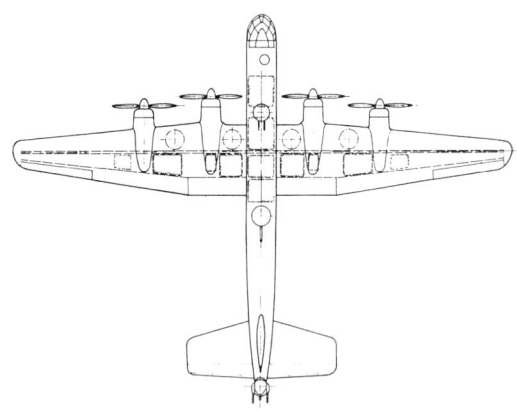

The He 177B-5 (Stage II) long-range bomber.

The range of all these proposed versions averaged around 4,400 km (2,734 miles), about 400 km (248 miles) less than that estimated for the production series He 177A-7. However, on 10 August 1943 it was ascertained that the first He 177A-7 would not be delivered in winter 1944 as expected but only in summer 1945. By contrast, according to the *Technische Amt* the He 177A-8 (basically an A-3/A-5 airframe with four individual DB 603s or Jumo 213s retrofitted) could be developed and brought up to production status much sooner. Not only that, but the A-8 would be about 60 km/h (37 mph) faster than the He 177A-7 as originally envisaged with two DB 610s.

The first A-8 pattern aircraft was expected on 1 January 1944, the first production series example in autumn 1944. In addition there was the He 177A-10 (an A-7 airframe with four individual engines), the first prototype of which was to be completed by 31 July 1944, although the first production series aircraft would not leave the assembly line before October 1945.

The A-8 and A-10 were subsequently redesignated He 177B-5 and B-7 respectively, the

The Long-Range Bomber Without a Chance   161

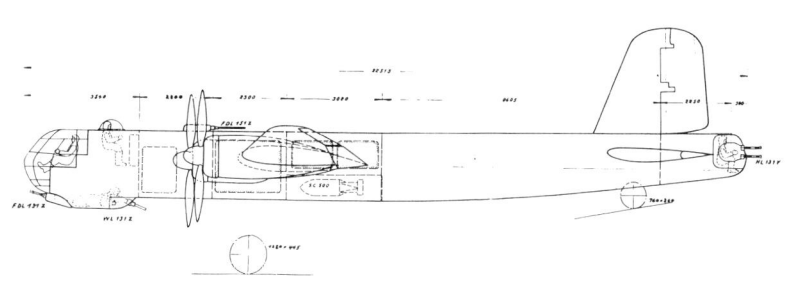

A three-view drawing from the He 177B-7 Type Sheet of 16 August 1943.

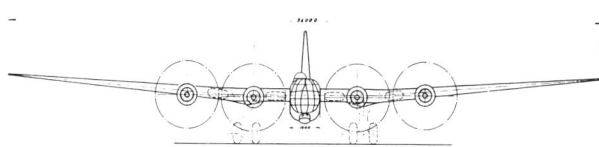

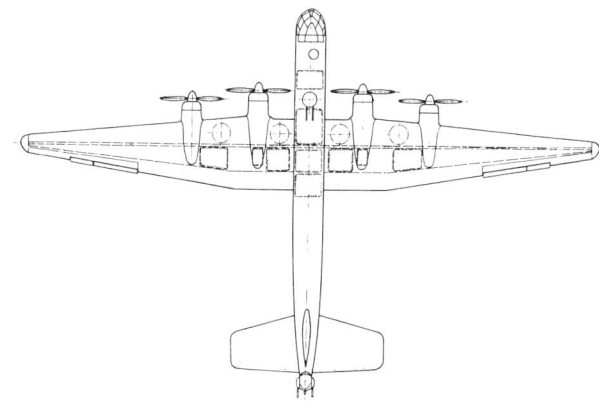

designations reflecting the A-5 and A-7 ancestry of the new versions. These designations first appeared in official documents in late summer 1943, by which time Heinkel's Development Department was ready with its timetable for the first He 177B-series aircraft:

Cockpit	He 177B-5 old	He 177B-5 new	He 177B-7 new
Construction data for pattern a/c	20 Oct 1943	15 Aug 1943	15 Nov 1943
First prototype	15 Feb 1944	01 Apr 1944	01 Aug 1944
Construction data for production series a/c	01 Mar 1944	01 Jun 1944	01 Oct 1944
First production series a/c	Autumn 1944	Mar 1945	Oct 1945

alte Ausführung

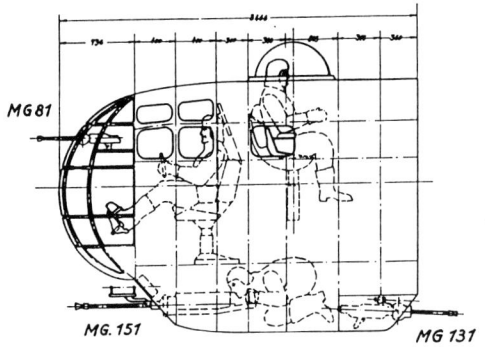
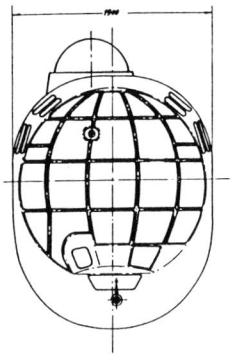

Comparison of old and new cockpit configurations as of 15 March 1943.

verbesserte Ausführung

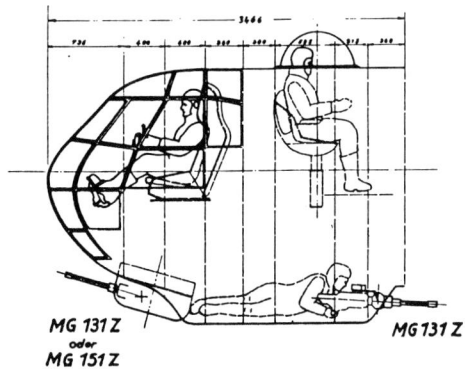
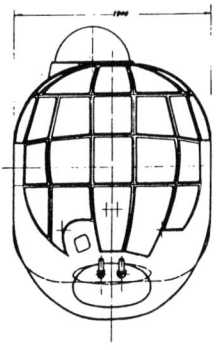

To enable the production of the He 177B-5 in some numbers as quickly as possible, it was suggested that a version featuring the old 102 m≈ (1,098 ft≈) wing area and a 31.40 m (103 ft 0 in) wing span be built.

For the Heinkel works, Director Francke proposed to drop the He 177A-7 (larger wing area, two DB 610s) as well as its four-engined counterpart, the B-7; production would thus be switched directly from the He 177A-5 to the four-engined B-5. To avoid any further delay, the new cockpit and twin fin/rudder tail assembly would also be deferred.

Although the mood of this *Generalluftzeug-meister-Amt* discussion on 10 August 1943 was really against the He 177, the final decision of *Generalfeldmarschall* Milch was clear:

'The He 177A-4 and A-5 will be produced as before. The He 177B-5 will be tackled with vigour. It will be built in series as soon as possible.'

Afterwards however, Milch gave permission to build only three B-5 pattern aircraft, and insisted that the A-5 be made operational at all costs.

For these reasons, in September 1943 the necessary development programme was given the highest priority. Naturally, the coupled engines were once again the focal point. The He 177A-3/V23, rebuilt as a pattern aircraft for the

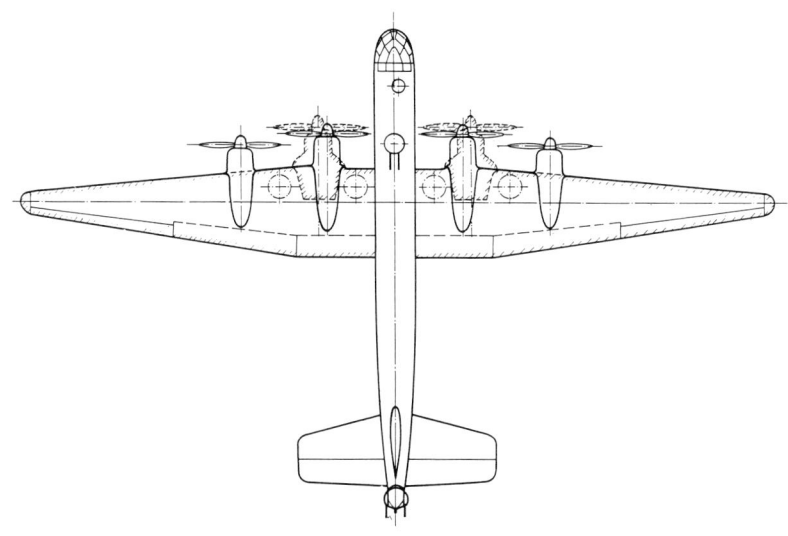

Comparison of the He 177A-7 (two coupled powerplants) and B-7 (four individual powerplants) dated 16 August 1943.

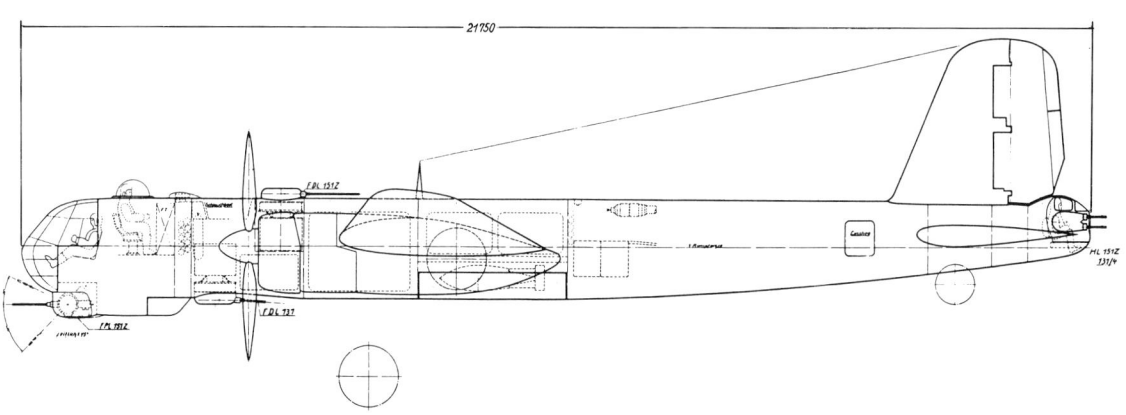

Side-view of the new fuselage layout envisaged for the He 177B powered by coupled DB 610s and with an HL 151Z tail turret.

A-5 series, was used to determine the He 177's flight characteristics and performance when powered by DB 610s repositioned 20 cm (7.87 in) further forward. The aircraft also featured an elongated fuselage and new slotted ailerons.

Another aircraft, the He 177A-0/V15, served as a test-bed at Zwölfaxing for the new cockpit and to investigate in-flight oscillations. These tests were to commence on 8 October 1943 and end with trials of the new wooden bomb-bay doors late in November 1943. At that time, there were another two prototypes at Heinkel-Süd's Vienna-Schwechat works: the He 177A-1/V25 (GI+BN/Wk-Nr 15153) pattern aircraft with large fin, and the He 177A-3/V34 (GP+WN/Wk-Nr 535364) fitted with a quadruple-barrel MG 131V tail turret.

Two more prototypes, the He 177A-08/V9 (GA+QQ/Wk-Nr 00 0023) and the He 177A-3/V101 (NN+QQ/Wk-Nr 535550), served as development aircraft for the different He 177B versions, carrying out many and varied tests. In summer 1943, the V9 (subsequently redesignated the V102) was withdrawn from the programme and fitted with a twin fin/rudder assembly. Test flights during November 1943 initially showed the new tailplane to be very satisfactory during take-off and to offer a more positive feel than

The HD 151/1 and HD 151/2 gun turrets were intended for the He 177B-5 and B-7 respectively.

with the single central tail. On the negative side, the effect of the new twin fin/rudder was considered to be somewhat lacking when the aircraft was coming in to land with the Fowler flaps down.

The first pattern aircraft fitted with four separate DB 603s, the V101, was cleared for flying on 5 January 1944. Initial taxying trials and the localisation of possible ground loop tendencies were soon followed by static thrust tests and the first stability flights.

In the meantime, tests had also been completed with the He 177A-5 fitted with the improved and repositioned DB 610 coupled powerplants. Early in October 1943, *E-Stelle* Rechlin had reached the conclusion that from the powerplant point of view the He 177B-5 was ready for operational service. The He 177A-3/V19 (VF+QA/Wk-Nr 332101), fitted with modernised DB 610s, had reported a good 50 flying hours at increased power, and only then suffered overheated pistons. On one occasion, the DB 610s fitted on the He 177A-3/V24 (ND+SS/Wk-Nr 135024) even managed 100 hours of flying without a single breakdown.

There were also other improvements, such as the disposal of fuel and oil leaks within both powerplants, which considerably increased flight safety; despite the still frequent oil spillages there were no more engine fires. The same applied to the flame dampers and oil coolers. It is of interest to look at the various engine problems in detail, and to note that the following faults had led to many fires involving DB 606 engines:

- Broken connecting rods in engines
- Oil and fuel leaks
- Insufficient access to the engine installation
- Inadequate servicing/maintenance
- Chaffing of fuel, oil and hydraulic piping
- Insufficient disposal of leaked oil and fuel

By the end of 1943 the Heinkel works had succeeded in eliminating partly or completely many of these serious defects. This was mainly thanks to comprehensive improvements in the powerplant sphere:

- Powerplants moved forward by 20 cm (7.87 in)
- Modification of powerplant carriers and their lateral supports
- Diversion of oil and fuel leaks via sheet metal
- Fitting of exhaust gas outlets behind the firewall
- Improved ventilation of engine compartment
- Fitting of enlarged oil coolers
- Installation of more powerful oil centrifuges

The operational safety of DB 610 engines was also the main subject of the *Generalluftzeugmeister-Amt* conference on 30 November 1943. The RLM representatives seemed very optimistic and expressed the hope that engine fires had now been completely eradicated and an unhindered start could be made to large-scale series production. *Generalfeldmarschall* Milch even expressed the view that if the engine situation remained as it was, he could see no reason to switch He 177 production to the He 177B-5/B-7. Nevertheless, work continued on aircraft with four individual engines so that the first so-configured example could leave the assembly line in October 1944, as planned.

The development discussion on 17 December 1943, chaired by *Generalfeldmarschall* Milch, began with a comparison between the He 177 and the Ju

288. The discussion turned once again to the difficulties experienced by the *Luftwaffe* during bombing operations in the West. Dipl-Ing Heinrich Hertel, who had left his post as Heinkel's Technical Director and Chief of Development in early 1939 to work as Junkers' Technical Director, and who oversaw development of the Ju 288 as the most promising of the 'Bomber B' contenders, duly responded with another attempt to boost the chances of his favourite. He argued that the Ju 288 was, on paper, 90 km/h (56 mph) faster than the He 177B-5, 50 km/h (31 mph) more than expected by *Oberst* Petersen (KdE). *Generalmajor* Pelz, too, supported keeping the Ju 288 in the production programme, at least because of its speed advantage. However, *Oberstleutnant* Knemeyer (GL C/E) immediately advised all present to give up on the Ju 288 because its operational capabilities would be covered by the Ju 488 in terms of altitude and the Jumo 222-powered Ju 388 in terms of speed.

Apart from that, various forms of offensive action, even in the West, were offered by the still-available He 177s on the one hand and such modern developments as the Do 335 *Pfeil* and Ar 234 *Blitz* high-speed bombers on the other.

In the meantime, completely untouched by the wrangling between the RLM and OKL, Göring and Milch, and Heinkel and Daimler-Benz, trials in support of the He 177B-5 continued at Heidfeld and Zwölfaxing with a reasonable measure of success. The first provisional B-5 powered by four individual engines made a successful first flight on 20 December 1943; the He 177 V101 followed on 3 January 1944. In strong squalls, the long-range bomber achieved 448 km/h (278 mph) near to ground level, but due to a sudden failure of the hydraulic system the test flight had to be terminated prematurely. To give added momentum to the He 177B-5 test programme, the RLM was then asked to allocate another two He 177A-5s from the current production.

But all was not as well as it seemed, for serious problems concerning series production of the He 177B-5 had already manifested themselves at the Arado works. It had been planned to complete the first production-standard B-5 there in October 1944, but the quick start to series production of the Ar 234 resulted in an inevitable delay to the B-5 schedule of one month.

It had become obvious already in early 1944 that adhering to any given timetables for development and production of the B-5 was all but impossible, even when it came to the construction of pattern aircraft. Zwölfaxing had hangar capacity for just six prototypes, and thus only two of the four He 177B-series aircraft, the V102 and the V103 (KM+TL/Wk-Nr 550036), could have their own parking area. For the lack of maintenance capacity an important series of tests, such as in-flight oscillation experiments with the He 177 V15, had to be cut short on 7 February 1944.

Nevertheless, the work programme drawn up

The only known photograph of the He 177B-5/V101 (NN+QQ/Wk-Nr 535550), which made its first flight on 20 December 1943.

by Heinkel-Süd's Vienna-Schwechat plant in February 1944 showed clearly that Heinkel had no intention of letting development of the He 177B-5 remain static. The programme traces stage-by-stage the planned evolution of a long-range bomber with an all-up weight of 44,000 kg (97,003 lb) and a range of 8,000 km (4,971 miles):

Stage 1: Two manually-operated MG 131s in nose gun position of A-5, plus retrofitting of A-3s by external technical service and service personnel

Stage 2: Introduction of twin fin/rudder assembly beginning with He 177B-5 series. Testing of new tailplane to be completed mid-April 1944.

Stage 3: Strengthening of defensive armament with remote-controlled MG 131Z in nose gun position of He 177B-5, and upgunning of both dorsal positions with MG 151Zs.

Stage 4: Application of reworked cockpit with one MG 151 in nose gun position, replaced by MG 151Z or MG 131Z in lower nose gun turret by autumn 1945.

Stage 5: Extension of wing span to 36 m (118 ft), increased wing area as on the He 177A-7, four separate engines, increased fuel capacity, strengthened tailwheel and jettisonable auxiliary support undercarriage assemblies.

Stage 6: Progressive development of He 177 airframe combined with Me 264 wing; wooden twin fin/rudder assembly and steel fuselage structure.

Some of these stage-by-stage improvements were supposed to be undergoing their preliminary practical clearance tests already on the first four He 177B-series prototypes.

While the He 177 V101 still had the regular central fin/rudder with paddles, the V102 was intended to have a twin fin/rudder assembly from the outset. Initially, the V103 was to have the single fin/rudder, which was then to be replaced by a modified twin fin/rudder assembly early in July 1944.

On 24 February 1944, the first two He 177B-series prototypes were inspected at Vienna-Schwechat by *Oberstleutnant* Knemeyer

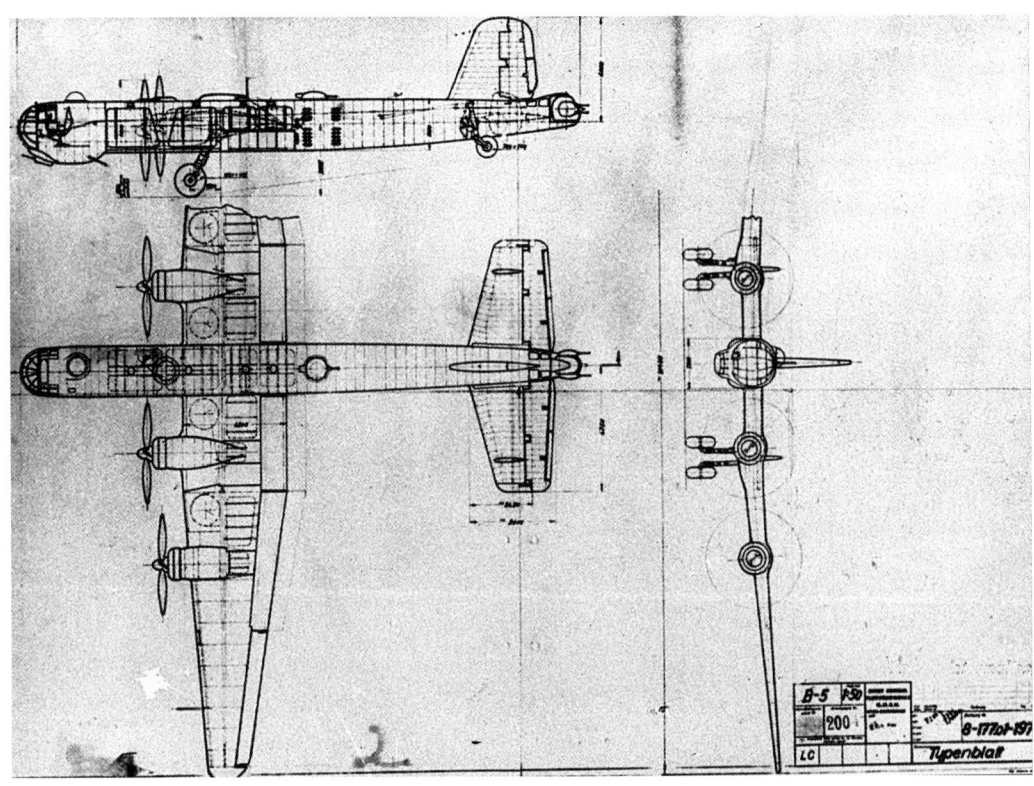

Three-view drawing of the He 177B-5 da 7 February 19 showing differ powerplant configurations

THE LONG-RANGE BOMBER WITHOUT A CHANCE   167

(GL C/E), *Oberst* Petersen (KdE) and *Oberst* Marienfeld (GdK). Shortly afterwards, *Generalfeldmarschall* Milch arrived at the Vienna-Neustadt airfield to gain his own impression of progress of the B-series development programme to date.

The first He 177 with a twin fin/rudder assembly was then test-flown in succession by *Oberst* Petersen, *Oberstleutnant* Knemeyer, *Hauptmann* Meyer and *Hauptmann* Fritz Schäfer. Knemeyer, who was piloting the He 177 for the first time, was enthusiastic; he had never believed that a four-engined heavy bomber could have such excellent handling qualities. As a result, Knemeyer expressed his belief that development of the He 177B-7 should be considered even more urgent than the Heinkel P 1068 four-engined jet bomber project (essentially a scaled-up Ar 234 with an anticipated bomb-load of 2,000 kg (4,409 lb) and a range of 1,600 km (994 miles), later designated the He *Strahlbomber* 16t, and then the He 343, but never flown).

A comparison fly-off between the He 177B-5 prototype and the KdE's touring aircraft, a Ju 188, showed the Heinkel bomber to possess a marked superiority in speed. *Oberst* Petersen then decided that all four B-5 prototypes should complete their factory flight tests at Vienna-Schwechat as quickly as possible before being transferred to *E-Stelle* Rechlin. The following tasks had to be completed first:

V101: Flight with two 'dead' engines on one side, using paddles.
Performance and speed test flights
Stability at extreme cg positions
Operational ceiling on two/three engines
V102: Tests of twin fin/rudder assembly
Tests of new powerplant installation
Performance measurements
V103: Inclined flights up to 700 km/h (435 mph)
Overload take-offs at an all-up weight in excess of 33,500 kg (73,854 lb)
V104: Production series prototype

A little later, all details pertaining to the planned He 177B-5 series production were cleared during a detailed discussion. Production was to commence using DB 603A engines, these to be replaced from July 1944 onwards by the more powerful DB 603E.

There were quite considerable changes in the defensive armament of both the B-5 and the B-7. Instead of the remote-controlled twin-barrel MG 151Z turret, beginning in March 1945 the production series aircraft were to be fitted with a modified HL 131V tail turret fitted with four MG 131s. A similar turret had already been built by Junkers for the Ju 290B-1, and could thus serve as a pattern.

Production of the He 177B-5 was to start in autumn 1944, commencing with five pre-production aircraft for which Heinkel had already received an official order. Compared to the He 177B-5, the B-7 featured several important differences:

- He 177A-7 wings with four individual DB 603E engines

The DB 603 engine, planned to power several He 177B-series models.

It was intended to adopt various nose and tail turrets developed for the Ju 290, such as the twin-barrel FDL 151Z (above) and the quadruple-barrel HL 131V (right), without fundamental alteration.

- Additional jettisonable auxiliary support undercarriage assemblies
- Strengthened tailwheel installation with a larger wheel
- Increased fuel tankage
- Improved cockpit with better visibility and strengthened defensive armament comprising:
  Nose: BL 131V with 500 rounds per barrel
  Dorsal (fwd): FDL 151Z with 500 rounds per barrel
  Dorsal (aft): 1) deleted on production series aircraft
  2) HD 151Z with 500 rounds per barrel
  3) HD 151/2 with 500 rounds
  Ventral: FDL 151Z with 300 rounds per barrel
  Tail: HL 131V with 1,000 rounds per barrel

It has to be noted that the ventral defensive position was once again to be taken unchanged from the Ju 290. The same applied to the undercarriage, thus enabling the first He 177B-7 to be tested as early as February 1945. Heinkel intended to manufacture most of the sub-assemblies in France, with final assembly and flight-testing to take place in the less dangerous Vienna area.

But none of these plans came to fruition. For

The Long-Range Bomber Without a Chance 169

Another view of the quadruple-barrel HL 131V tail turret mock-up.

capacity reasons, the entire He 177B-series programme had to be considerably scaled back in favour of the He 274 development programme.

Tests and Trials to the Last Day

Despite a lack of workshop capacity, Heinkel managed to keep two of the four-engined B-series aircraft in a constant state of airworthiness. Using the V101, it was established that, thanks to the wing spoilers and the paddles fitted to the horizontal tailplanes, it was easy to carry out flights using only the engines on one side of the aircraft at speeds up to 220 km/h (137 mph). Further tests proved that all take-offs at an all-up weight of about 31,000 kg (68,343 lb) could be accomplished without the tendency for the aircraft to ground loop as feared by the RLM. The aircraft's climbing performance also corresponded to all figures guaranteed by the manufacturer.

Prior to these tests, on 13 May 1944, the Heinkel-Süd works had been visited by General Peltz. During a frank discussion with Prof Dr-Ing Heinkel and his most important colleagues, the GdK stated that:

> '. . . because of the high loss rate caused by powerplant defects, as well as the resulting low operational readiness, the current He 177 with its coupled engines cannot be tolerated much longer.'

Because of the complexity of the He 177, workshops located on distant airfields were faced with problems that were all but impossible to solve, which led in turn to protracted out-of-commission times for individual aircraft. It was often months before the required spare parts could be obtained; in the meantime, it was inevitable that short-circuiting would occur in the electrical systems of He 177s parked of necessity in the open for long periods.

However, General Peltz also emphasised that basically he had nothing to criticise regarding the He 177 airframe; there was something wrong with the powerplants. In his opinion, the He 177 with coupled engines would have been deleted from production immediately, had it not been for the prospect of the B-5. Until then, Prof Dr-Ing Heinkel and his team had to do all they could to improve the serviceability rate of the He 177A-3 and A-5 by means of a considerably enhanced external technical service.

Further development of the He 177B-5 was accorded great importance in the revised flying programme of 28 June 1944. By then, flight tests with the V101 had revealed that the aircraft began losing stability at higher speeds. At around 400 km/h (248 mph), the test pilots reported an almost complete loss of stability, especially with the aircraft at an average all-up weight. Modifications were urgently required.

The He 177B-5 V102 had been transferred to *E-Stelle* Rechlin on 14 March 1944 to complete the powerplant and wing de-icing equipment tests. During a continuous engine test flight on 13 April the crew were informed of an MYO* warning and landed on the nearest airfield,

---

* *Luftwaffe* service warning of possible air raid within 15 minutes, necessitating immediate landing of all aircraft in the vicinity except fighters under orders to intercept.

Heidfeld-Schwechat. After touchdown the aircraft went into a skid and was badly damaged as a result. This meant that the second experimental He 177B-5, fitted with a twin fin/rudder assembly, was now temporarily out of action: the repairs were not expected to be completed before 25 June 1944. To save time, the repair period was used to re-engine the aircraft with four DB 603Es, so that testing could recommence at *E-Stelle* Rechlin immediately after the airframe repairs had been completed. However, it is not known if this was ever achieved; the aircraft was still parked at Vienna-Schwechat early in February 1945.

The He 177B-5 V103, a converted He 177A-5, was to have a brake parachute fitted by 15 May 1944. The aircraft made its first and second flights on 26 June 1944, these being used to confirm the B-5's loss of stability at higher speeds. In contrast, inclined flight tests provided proof of the airframe's strength at a maximum speed of 700 km/h (435 mph).

The He 177B-5 V104 (KM+TE/Wk-Nr 550005), another rebuilt A-5, was intended as a production prototype and should have started flight tests on 15 June 1944 fitted with quadruple-barrel nose and tail turrets as well as a twin fin/rudder assembly. As it happened, there were personnel shortages late in the month of May which made it impossible to set a definite completion date for the aircraft. For one thing, installation of the new nose gun position created numerous detailed technical problems, although subsequent tests with the modified defensive armament installation proceeded smoothly at *E-Stelle* Tarnewitz and were completed by 15 July 1944.

But there were also other obstacles that were non-technical in nature, such as difficulties in material acquisition which increasingly slowed down the entire He 177B-5 flight test programme. Enemy heavy bombing raids did the rest for from spring 1944 onwards, the Heinkel-Süd works were targeted several times by Allied bombers. A raid by 192 B-24 Liberators of the US 15th Air Force was intended for the Vienna-Schwechat site, but hit only the residential buildings in the Schwechat and Zwölfaxing municipalities. The company's luck also held on 8 April 1944 when USAAF B-17 Flying Fortress crews were unable to locate their targets due to the foggy weather.

It was a different story on 23 April, when the USAAF made a concentrated effort to hit the German aircraft industry in the Vienna area. A force of 956 mostly heavy bombers crossed the Alps and homed in on the Heinkel-Süd works. Within 15 minutes 265 tonnes (292 tons) of bombs fell on the Heidfeld-Schwechat plant, causing much damage and the deaths of 94 workers. A month later, on 24 May 1944, bombs rained down on Zwölfaxing.

Following another bombing raid on the Heidfeld-Schwechat site, Heinkel had to terminate manufacture of the He 219 *Uhu* (Eagle Owl) night-fighter there. But worse was to come. After another visit by the bombers on 8 July things looked grim for construction of the He 177B-series pattern aircraft at Zwölfaxing: aircraft halls II and IV went up in flames, and the whole airfield was ploughed over by bomb craters. The V103 was destroyed and presumably also the V104, then under construction. As a result of the raid, Heinkel was forced to give up the Zwölfaxing works and concentrate its remaining production capacity in the similarly damaged Heidfeld-Schwechat night-fighter plant. By then, Prof Dr-Ing Heinkel had held intensive negotiations with the RLM regarding the forthcoming production of the He 177B-5 and B-7. Already in mid-May 1944 it was all but agreed to forgo the jettisonable auxiliary support undercarriage assemblies on the B-7 and to design a strengthened main undercarriage instead.

Another alteration discussed between Heinkel and the RLM was the necessity or otherwise of the quadruple-barrel nose turret for nocturnal operations. In this case, the GdK suggested that the company should build half the production run of bombers without the heavy nose turret, and possibly also without the aft dorsal gun installation. This would increase the maximum speed of the lightened He 177 to 569 km/h (354 mph). In addition, new considerations in June 1944 indicated that by reducing the defensive armament to just one MG 131Z each in the forward dorsal and tail positions and one MG 131I in the ventral position, it should be possible to extend the He 177's range to 7,000 km (4,350 miles).

The necessary additional fuel would be accommodated in the space released by deletion of the aft dorsal turret and inside tightly-riveted integral wing fuel tanks. More fuel would be carried in two 900-ltr (198 Imp gal) drop tanks. An even more drastic reduction of onboard equipment and the abandonment of the oxygen installation would, it was calculated, reduce the all-up weight to 40,000 kg (88,185 lb) and thus boost maximum range to 7,500-8,000 km (4,660-4,971 miles).

On 8 June 1944, attention turned once more to the question of reduced defensive armament. It had been calculated that the range of an He 177 equipped with two 1,200-ltr (264 Imp gal) drop tanks would be extended from 5,000 km (3,107 miles) to 5,800 km (3,604 miles). However, an He 177 armed only with quadruple-barrel nose and tail turrets would have a take-off weight of 36,000 kg (79,366 lb) and a maximum range of 5,700 km (3,542 miles); but if fitted with two 1,200-ltr (264 Imp gal) drop tanks, its range would increase to at least 6,500 km (4,039 miles) – a welcome increase, but still far short of the initially estimated range of 8,000 km (4,971 miles) which, it was realised, could not be achieved.

Apart from the He 177B-5 and B-7 both the Heinkel design office and the *Technische Amt* of the RLM were busy with some additional experimental aircraft intended to serve as flying test-beds for various improvements for the He 177A-5.

One of these aircraft was the V15 which, after the appearance of vertical in-flight oscillations in March 1944, was fitted with corresponding counterweights. A month later a Heinkel works crew achieved a maximum speed of 710 km/h (441 mph) in an inclined flight test in this aircraft. On safety grounds, the V15 was fitted with a large brake parachute, which functioned well on its first deployment. On the second flight, however, the parachute coupling rope was ripped apart, necessitating some strengthening of the complete installation. Unfortunately, on 24 June 1944 the V15 was badly damaged in a flying accident at Deutsch-Wagram; as a complete repair was considered uneconomical, the aircraft was released from trials work for cannibalisation for spare parts.

Another of the test aircraft was the V23. Following countless powerplant measurement flights with the repositioned DB 610 engines, not all of which produced satisfactory results, this aircraft was fitted with new de-icing equipment and improved ailerons. During subsequent test flights the aircraft ran into unexpected problems with the ignition of its heating system at high altitudes. Nevertheless, these tests were completed by May 1944, allowing the aircraft to be detailed the following month for the instruction of trainee aircrews. Unfortunately the aircraft was slightly damaged during an enemy low-level attack during June, and on the 28th of that same month was reserved for transfer to FFS (B) 31 at Brandis.

The V25, the pattern aircraft for the enlarged single tailfin, also tested enlarged horizontal tailplanes. In June 1944, these tests were continued by the He 177A-5/V39 (KM+TY/Wk-Nr 550049).

Another aircraft worth mentioning is the He 177A-3/V40 (GR+MH/Wk-Nr 535850), which was used as the prototype for the pattern installations of the FuG 217 and *Berlin* radar devices. After repairs of Fuselage Frame 41, cracked during flight-testing, the aircraft was to be fitted with both radar devices by early June 1944. However, completion of this work was delayed until the end of July at the earliest by a lack of technical personnel. Later, this aircraft was delivered to the *Luftwaffe* for use at *E-Stelle* Werneuchen.

Even on 25 May 1944 it seemed certain that the He 177 would remain an important element in the official development and production programme. During a conference at Obersalzberg on that day when those taking part included *Reichsmarschall* Göring, *Generalfeldmarschall* Milch, *Generalleutnant* Galland (GdJ), General Vorwald (GL/C), General Korten (Chief of the *Luftwaffengeneralstab*), *Oberstleutnant* Petersen (KdE) and *Oberst* Knemeyer (GL C/E); as well as Party Leader Otto Sauer (in charge of the *Jägerstab*), Dr Frydag (Director of the Main Committee for Airframes and former General-Manager of Heinkel), *Fliegerhaupt*-Ing Reidenbach, Prof Dr-Ing Heinkel, Prof Dr-Ing Willy Messerschmitt, Dipl-Ing Hertel and Dipl-Ing Kurt Tank, it was decided to delete from the production programme not only the Ju 52, Ju 188, Ju 288 and Ju 390 but also the Bf 110 and He 219 *Uhu* – and that these cuts would be followed by the deletion of many other aircraft types.

After phasing out of production the four-engined Ju 290 and six-engined Ju 390, the He 177 would remain as the only heavy bomber in production for the *Luftwaffe*; as such, it was expected to operate in numerous other operational roles as well, such as torpedo-bomber, very long-range reconnaissance aircraft and the so-called '*Amerika Bomber*'. With this in mind, Göring even increased the planned output of the He 177B-5 to 200 aircraft per month!

As a result of the decision to cancel further production of the He 177 in favour of fighters and fighter-bombers, the main beneficiary was the turbojet-powered Me 262.

Along with the Me 262A-1/-2, production of the Do 335 Pfeil was to have been accelerated at the expense of the He 177 and He 277. Illustrated is the Do 335 V9, assigned to E-Stelle Rechlin in May 1944.

It therefore came as something of a surprise to the Heinkel works when, just a month later, they received a new directive from the Minister for Armaments and War Production, Prof Albert Speer. Following an agreement with *Reichsmarschall* Göring, the *Jägerstab* had issued an order, with Hitler's approval, stating that the main emphasis in aircraft production should be on fighters – and that further production of heavy bombers should be deleted completely.

This was followed by an instruction from Göring requesting a speedy phasing out of production of the He 177A-5 and ordering the cancellation of the entire He 177B-5 and B-7 development programme forthwith. All personnel hitherto engaged on these tasks were to be allocated to development and production of the He 162 *Volksjäger* (People's Fighter) single-seat turbojet-powered interceptor, otherwise known as the Salamander, or to work involving

Following the discontinuation of all further German heavy bomber development, extra production capacity was allocated to the He 162 Volksjäger.

the He 219 *Uhu* night-fighter and new jet aircraft development programmes.

Final Employment: The Mistel

To find an operational use for the remaining 200 or so intact He 177s parked on various airfields in Germany, Denmark and Norway, the Heinkel works proposed to modify them as the lower elements of *Mistel* (Mistletoe) composites, similar to Ju 88s already so modified, and thus make them available for attacks against selected important targets.

Originally proposed and rejected as a viable weapon in 1941, the *Mistel* concept was revived in early 1943. The composite comprised a war-weary Ju 88A-4 with its forward fuselage replaced by a large hollow-charge warhead to produce a twin-engined missile flown to the target by a piloted single-engined fighter mounted atop, 'piggy-back' style by means of a steel-tube structure linking the wing main spars of the two aircraft. A single strut to the rear supported the fighter's tail; when the pilot had lined up the composite with the target he released the rear strut which fell backwards into a yoke on the rear fuselage of the Ju 88, triggering an electrical contact which in turn released the main attachment points on the steel-tube tripods and let the fighter fly free. The Ju 88 missile would then continue on its set course towards the target while the fighter climbed away.

Codenamed *Beethoven*, the *Mistel* concept flew in prototype form in July 1943, a Bf 109F-4 being mounted atop a Ju 88A-4. This combination became the *Mistel* 1 and was joined in due course by the *Mistel* 2 (Ju 88G-1 and Fw 190A-6) and *Mistel* 3C (Ju 88G-10 or H-6 and Fw 190A-8); training versions comprised the *Mistel* S1 (as per *Mistel* 1), the S2 (Ju 88G-1 and Fw 190A-8) and S3A (Ju 88A-6 and Fw 190A-6).

The operational debut of the *Mistel* composite came in June 1944, when *Mistel* 1s assigned to the *Einsatzstaffel* (Combat Operations Squadron) of IV/KG 101, which had converted to the type during spring 1944, attacked Allied shipping in Seine Bay. The attack met with qualified success: although the Ju 88A-4 missiles reached their targets, none exploded on impact and the vessels thus remained afloat. Later that year, in October, the unit formed the nucleus of what was intended to be a dedicated *Mistel*-equipped unit, the *Einsatzgruppe* (Combat Operations Wing) of

174 Heinkel He 177-277-274

A Mistel 2 composite assigned to 2./KG 30 'Adler' at Oranienburg in 1945. Note the large warhead in place of the Ju 88G-1's forward fuselage and the auxiliary fuel tanks carried beneath both aircraft.

A Mistel S3 training composite at Stassfurt in April 1945. The He 177/Fw 190 composite would have looked still bigger next to it.

III/KG 66. However, just days later this unit was redesignated II/KG 200.

During summer 1944, the commander of KG 200 had obligingly informed the Chief of IV *Fliegerkorps* that the *Mistel* 3C composite had an operational range of only 1,500 km (932 miles), but that an He 177 combined with an Fw 190A-8 and armed with a special large mine charge would have a range of 3,300 km (2,050 miles). Based on this estimate, such a composite would have a penetration depth of 1,650 km (1,025 miles) – thus bringing 10 of the most important power stations in the Soviet Union with a combined output of 220,000-600,000 kW into range.

Taking off from Prowehren in East Prussia, the He 177/Fw 190 *Misteln* would have to cover a distance of 1,100-1,250 km (683-777 miles) to reach these targets. A planning study envisaged the destruction of all 10 targets by a force of 28 He 177 *Misteln*. In case the *Misteln* were unavailable, *Generaloberst* Ritter von Greim wanted to attack the targets with volunteer 'self-sacrifice' pilots in what was now codenamed Operation *Eisenhammer* (Iron Cross).

On 20 November 1944, Director Francke of Heinkel presented the finalised plan of the He 177/Fw 190 *Mistel* composite to General-Ing Hermann (OKL Fl-E) and proposed the construction of one experimental aircraft. According to Francke, a second prototype could be ready any time within 15 days. On a larger scale, for the conversion and construction of 50 He 177 *Misteln* Heinkel estimated it would take about 5,000 manhours per airframe, the work to be carried out at the Eger plant.

On 17 December *Luftflotte* 6 requested the soonest possible availability of 50 He 177 *Misteln* for operations on the Eastern Front. But by now certain doubts had arisen concerning the carrying capacity of the He 177 as the lower component of the *Mistel* composite, thus casting doubt on all of the planning already completed. Nevertheless, preparations for Operation *Eisenhammer* went ahead, and at the end of 1944 most of the He 177A-3 and A-5 aircraft parked on several airfields were brought up to full flying condition once again. According to clear and specific instructions from the OKL, the attack on the Soviet power stations had to be carried out during the next full moon period, in other words at the end of January 1945,

A binding contract to re-equip 50 He 177A-3/A-5s for use in Operation *Eisenhammer* was received from the RLM early in December 1944. A short while later the total number of aircraft involved was reduced from 50 to 20, so as not to endanger production of the He 162 *Volksjäger* fighter. Production of even this reduced quantity of He 177 *Misteln* was, however, unachievable. By 7 December 1944 neither of the two prototypes on order was ready, and due to the ravages of war no pattern aircraft could be completed early in 1945 either. For this reason the Chief of *Technischer Luftrüstung* (Technical Air Armament) ordered in January 1945 the transfer of two He 177s to Nordhausen, where they would be converted under his own direction.

It seems that a few He 177s were actually converted as *Mistel* composites, possibly by *Luftwaffe* ground personnel. On 20 February 1945 RAF reconnaissance aircraft took a series of aerial photographs of the Grove and Tilstrup airfields in Denmark, and after careful scrutiny the photo-interpretation personnel positively identified 12 of the 35 well-camouflaged *Mistel* composites parked on these airfields. Most of those identified were Ju 88/Fw 190 combinations, but a few He 177/Fw 190 composites were also noted. A low-level attack ordered by RAF Fighter Command shortly afterwards destroyed or damaged most of these *Misteln*. Examination of post-strike reconnaissance photos led the RAF experts to believe they had identified one He 177 '*Mistel* Monster', as it was dubbed.

Exactly how many He 177s were converted as the lower components of *Mistel* composites is unknown. It is certain however, that no He 177 *Misteln* were ever used in action; the only 'piggy-back' unmanned bombers used operationally were the Ju 88s.

Chapter 5

The He 274 and He 277 Strategic Bombers

The He 177 High-Altitude Bomber: A Wasted Chance

Even before the start of the Second World War, consideration was already being given to the use of the future He 177 as a high-altitude bomber. The first draft design of a three-seat pressurised cockpit for a high-altitude version of the He 177 was submitted to Prof Dr-Ing Heinkel for his expert appraisal on 27 April 1939. Just as important was the design of pressurised weapon installations and the layout of the tail gun position as a special pressurised cabin in its own right.

More than 18 months went by before reliable information was at hand, with the result that the idea of an He 177 with a pressurised cockpit made another appearance in the development review of 11 December 1940. This long delay also had an adverse effect on development of the projected He 177A-2 high-altitude bomber equipped with a pressurised cockpit.

Heinkel's Development Department envisaged the use of the He 177A-08, A-09 and A-010 pre-production series aircraft as the V9, V10 and V11 prototypes respectively, although the first pattern aircraft with a pressurised cockpit was not expected until summer 1941. The second high-altitude prototype, equipped with a long-range reconnaissance installation, was scheduled for completion in August 1941, and was to commence factory flight tests that same month. The pattern aircraft fitted with double- and triple-barrel gun

Development of the He 177H's pressurised cockpit drew on practical experience gained with the two Junkers EF 61 high-altitude research aircraft. Illustrated is the EF 61 V1, destroyed in a crash on 19 September 1937.

turrets as well as a pressurised cabin was to be the V11, which Heinkel intended to complete by September 1941 at the latest.

In February 1941, with both the He 177 V2 and V3 prototypes having been lost in crashes, only the V1 remained in flying condition, and consequently all previously planned 'cleared for flying' dates for the pressurised cockpit-equipped aircraft had to be postponed until November 1941. Work on the projected V9, V10 and V11 equipped with pressurised cockpits received a further setback in mid-1941 with the loss of the He 177A-01, leaving at most three prototypes to carry on the He 177 flight test programme. In mid-August, by which time both the V9 and V10 were originally supposed to have been completed and starting flight-testing, only the He 177A-02 to A-07 and A-011 to A-015 were in the final stages of assembly at Rostock-Marienehe. At the same time, the first He 177A-0 pre-production series aircraft produced under licence by ARB and HWO were receiving their finishing touches.

Extensive and comprehensive ground tests to ensure the continuous seal-proof condition of the pressurised cockpit were finally completed on 20 August 1941, enabling the first provisional pressurised cockpit for the He 177A-4 high-altitude bomber to be on hand at the end of that month. Coinciding with the planned fitting of the He 177A-015 (GA+QX/Wk-Nr 00 0030) with four DB 610F high-altitude engines, Prof Dr-Ing Heinkel also submitted a proposal for three additional He 177 high-altitude bomber projects of differing configuration:

- He 177 with Junkers Jumo 208A diesel engines with superchargers
- He 177 with BMW 801Ks and multi-stage superchargers
- He 177 with DB 603As and TKL 15 superchargers

These projects, each of which envisaged a high-altitude bomber powered by four separate high-altitude powerplants, won unanimous approval during a long discussion between representatives of the Heinkel works and the *Luftwaffengeneralstab* on 10-11 October 1941. The projected aircraft, initially designated the He 77H but later redesignated the He 274, was designed to carry a 2,000 kg (4,409 lb) bomb-load over a range of 3,000 km (1,864 miles).

At the same time as the He 177H proposals were winning much favourable comment, the He 177A-4 powered by two DB 606 coupled powerplants and equipped with a pressurised cockpit was also attracting increasing interest. For the time being though, the number of prototypes with pressurised cockpits had to be temporarily reduced from three to two, because the A-08/V9 (GA+QQ/Wk-Nr 00 0023) was needed for flight-testing of the mechanical flight controls. By now the A-09/V10 (GA+QR/Wk-Nr 00 0024) and A-010/V11 (GA+QS/Wk-Nr 00 0025) prototypes were scheduled to make their first flights on 25 December 1941 and 25 January 1942 respectively.

By mid-November 1941 the first high-altitude bomber prototype was in the final stages of assembly. Work was also progressing well on the V11. However, the delivery of both aircraft had to be postponed due to a series of tests of the revised hydraulic system and the need to incorporate important airframe strengthening.

As a result, the V10 did not make its maiden flight until 21 January 1942. A short while later, according to planning, the V9 was to be transferred to Heinkel-Süd's Vienna-Schwechat works. There, the re-equipment of the provisional pressurised cockpit as well as the fitting of comprehensive measuring equipment was considered to be of the utmost importance. Yet some nine months later, in September 1942, the first of the two aircraft with pressurised cabins was still parked on the Rostock-Marienehe works airfield.

The He 177 V11 was not expected by the Rostock-Marienehe works management before 1 March 1942. It was to be equipped with a fully-functional pressurised cockpit which was expected to be available before the BAL acceptance date. In the meantime, an order had arrived from the *Technische Amt* requesting that the He 177A-4 be fitted with a pressurised manned tail gun position in addition to the pressurised cockpit. Accordingly, a single-seat functional mock-up was built early in 1942.

For a while progress of the test programme seemed so promising that in March 1942 it was even briefly planned to drop the He 177A-3 and produce the A-4 high-altitude bomber on a large scale instead. However, by summer 1942 it was evident that the A-4's test programme would take quite some time to complete. For this reason the production plan of 15 July 1942 listed an initial 100 A-3s, to be followed by a large series of 2,128

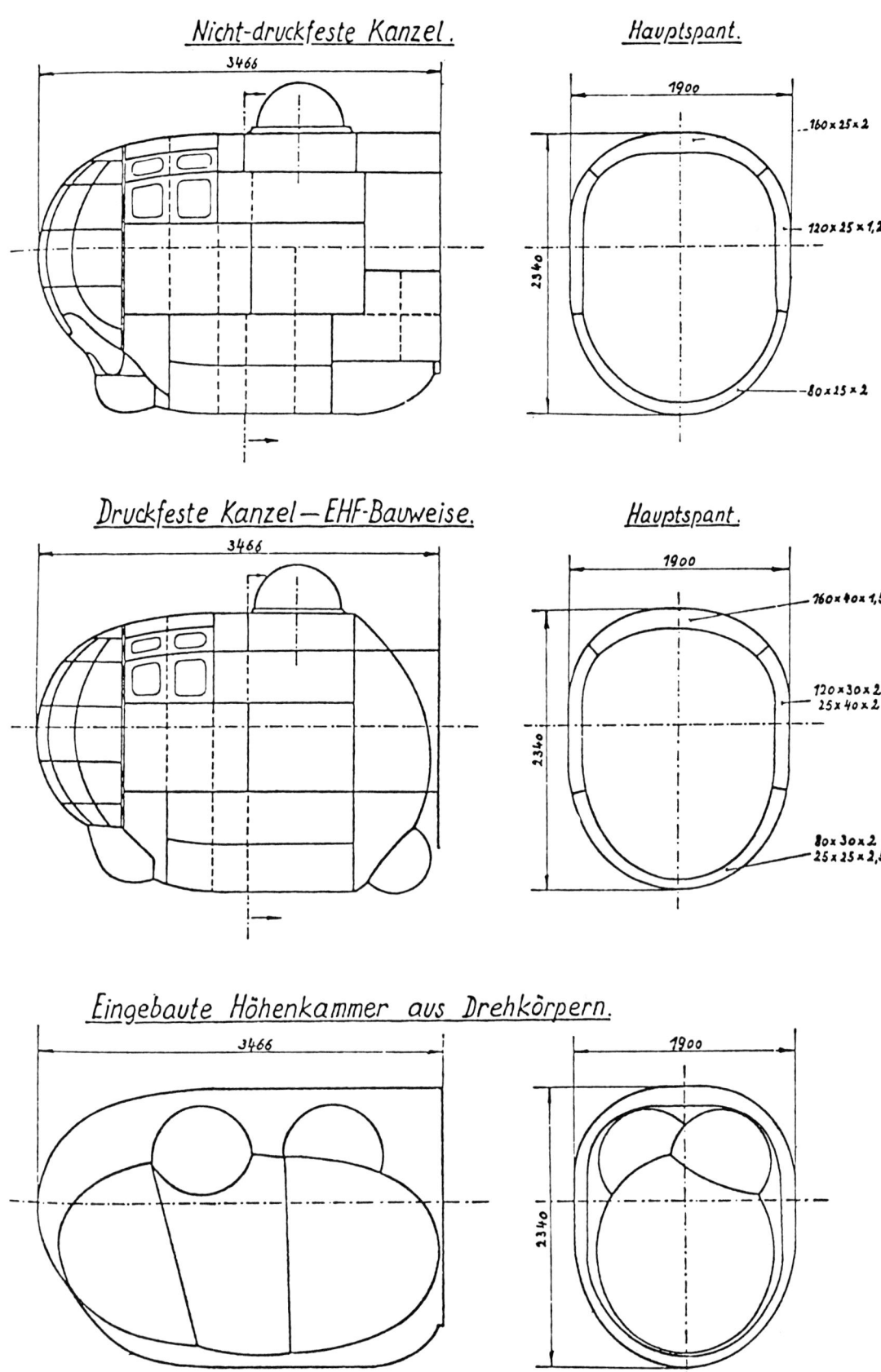

A comparison of the various pressurised cockpit forms planned for the He 177H/274 high-altitude bomber.

aircraft each fitted with a *Kutonase* (a barrage balloon cutter built into the wing leading-edge).

Alongside the proposed manufacture of the He 111R, an interim high-altitude bomber with reduced armour protection and defensive armament and powered by turbo-supercharged engines with annular radiators, there were also various projected high-altitude versions of the He 177 that shared the agenda during a conference held on 7 July 1942. One of the projects discussed was a new variant featuring a revised fuselage and a pressurised cockpit.

By February 1943, the list of all He 177 versions approved for further development included only the He 177A-5, A-6 and A-7, as well as the He 277. Development work on the two more complex high-altitude bombers, the A-2 and A-4, had been cancelled by the RLM shortly before, conversion costs having been deemed too high to justify large-scale production.

Trials with the pressurised cockpit had also come to a temporary halt when the He 177 V10 was badly damaged during testing, and until the aircraft was repaired and fitted with a new wing of increased span to help reduce wing loading and increase the aspect ratio, the entire experimental programme was to be carried out by the He 177 V11. By May 1943 repairs to the V10 were still unfinished in the Vienna-Schwechat works, and finally the V11 had to be temporarily grounded because its Daimler-Benz powerplants had to be changed once again.

With the demise of the He 177A-4, the V10 and V11 were eventually allocated as additional test aircraft for the He 274; the designation given to a high-altitude bomber development of the He 177 incorporating various new features deemed necessary as a result of the shortcomings in performance highlighted by the V10 and V11 during testing in support of the He 177A-4 programme.

On 3 August 1943 the V11 climbed to an altitude of 9,200 m (30,184 ft). During this test flight, only the pilot used an oxygen mask and onboard oxygen supply; the flight mechanic and the measurement engineer did not put their masks on. Despite the high altitude, all three crewmen felt quite well inside the aircraft's pressurised cockpit.

Later that same month there followed several sudden pressure drop tests from altitudes of around 7,000 m (22,966 ft). At the same time it was requested that the V10 be in full flying condition by 15 November 1943. The V11, meanwhile, was tasked with carrying out the following flight test projects:

- Investigation of windscreen ventilation
- Further development of cockpit heating
- Checking air supply by means of boost pressure

After this the tests with the pressurised cockpit were continued. On 12 October 1943 the flight engineer aboard the V11 opened the large pressure-outlet valve at an altitude of 7,000 m (22,966 ft). Theoretically, 65 seconds were needed to balance the pressure aboard the aircraft. During this time, the pressure ventilation was switched on, and the three-man crew survived the sudden pressure drop without any adverse physical effects. Apart from that, these tests with the V11, conducted up to 20 October 1943, revealed an inadequate tightness of the cockpit entry hatch.

Although quite a lot was achieved, taken as a whole it amounted to relatively few high-altitude test flights, during which the breathing air compressor also did not always function satisfactorily. There were other negative points: the pressure-holding valves proved to be usable only under certain conditions, despite several improvements, and nothing could be done to prevent the continuous loss of air pressure inside the cockpit. During another test flight to 7,000 m (22,966 ft) on 28 October 1943, an attempt was made to open the cockpit entry hatch at an inner pressure of 0.715 bar (i.e. air pressure at an altitude of 3,000 m/9,842 ft). This proved to be possible only after the application of great physical strength; the suggested solution was the fitting of a larger lever arm.

Early in March 1944, both the V10 and V11 finally had to be grounded due to a lack of workshop space and support personnel. By then, testing of the *Kehl*-equipped He 177 had assumed a higher priority; and the *Technische Amt* was more interested in the testing of increased defensive armament and the He 177 V101 to V104 prototypes.

The already strained situation had not improved by May 1944, and so the Heinkel works proposed that the V10 be cannibalised for spares and the V11 be transferred to *E-Stelle* Rechlin. In June the RLM issued an order transferring the

V10 to the Training/Instructional workshops at Rostock-Marienehe, while the V11 was transferred to Rechlin on 26 June to continue the pressurised cockpit tests. But it was only a short respite: following cancellation of the entire He 177 programme in July 1944 the V11 also lost its test aircraft status.

Development Task: 8-274

Following the demise of the He 177 V10 and V11 there was not a single He 177 fitted with a pressurised cockpit left in the experimental programme in late summer 1944 – a surprising state of affairs, given that development of the He 274 (formerly the He 177H) was about to enter its decisive phase.

The Heinkel design offices had begun work on the development of a dedicated high-altitude bomber version of the He 177 in 1939. The aircraft, dubbed the He 177H, was intended to operate at altitudes in the region of 15,000 m (49,213 ft) – safe from enemy anti-aircraft fire and fighter interception – and to carry a maximum bomb-load of 2,000 kg (4,409 lb). In the early stages of the project, no specific powerplant was selected; all power-boosted engines were under consideration, especially those with built-in turbo-superchargers.

During a discussion between Heinkel representatives and the RLM on 13 October 1941, it was agreed to build six experimental He 177Hs powered by high-altitude engines as soon as possible, each aircraft utilising an almost unchanged He 177A-3 fuselage. The Heinkel representatives had arrived for the discussion well-prepared and presented data on preliminary projects powered by different engines for comparison. The following facts were noted:

Because of a shortfall in design capacity at the Heinkel works, the RLM proposed to transfer construction of the aircraft, now formally designated the He 274, to factories in occupied France. Heinkel's design and performance estimates for the He 274 were then rechecked by *E-Stelle* Rechlin. However, Heinkel's proposal to construct the He 274's wings as steel spar structures soon had to be put on hold again; for in 1941, there were as yet no steel alloys that could withstand such extremely high demands in terms of transversal strength and load-carrying capacity.

A short while later, Heinkel received the official development contract for the He 274, which stipulated that construction of this new bomber would take place partly at Rostock-Marienehe and partly at the *Société Anonyme des Unises Farman* (SAUF) plant at Suresnes, near Paris. Two prototype and four pre-production series He 274s would be followed by the first production series aircraft. The He 274 was to have defensive armament comprising one 13 mm MG 131 machine-gun in the lower nose area and one double-barrel MG 131Z each in remote-controlled FDL 131Z forward dorsal and ventral barbettes. For a better field of fire, the ventral barbette was moved further aft, to a position behind the bomb-bay. Although the aircraft was to feature a twin fin/rudder assembly, there was no provision for tail armament.

The operational equipment and defensive armament configurations for the He 274 were fixed during a two-day conference at Rostock-Marienehe early in January 1942. All of Prof Dr-Ing Heinkel's proposals for the new high-altitude bomber were approved by the RLM, and the originally-planned armament was confirmed. A *Lotfe* 7D telescopic bomb-sight was to be fitted in the forward part of the cockpit; it would be operated by the navigator. The cockpit was to

	He 177	He 177H
Powerplants:	2 x DB 610F	2 x DB 603 fitted with DVL exhaust-driven superchargers
Take-off power:	1,350 hp each	2,000 hp each
Performance at altitude:	910 hp at 13,500 m (44,291 ft)	1,700 hp at 15,000 m (49,213 ft)
	750 hp at 15,000 m (49,213 ft)	850 hp at 17,000 m (55,774 ft)
Range with 2,000 kg (4,409 lb) bomb-load:	3,250 km (2,019 miles) at 550 km/h (342 mph) and 13,000 m (42,651 ft)	3,100 km (1,926 miles) at 615 km/h (382 mph) and 15,000 m (49,213 ft)

THE HE 274 AND HE 277 STRATEGIC BOMBERS   181

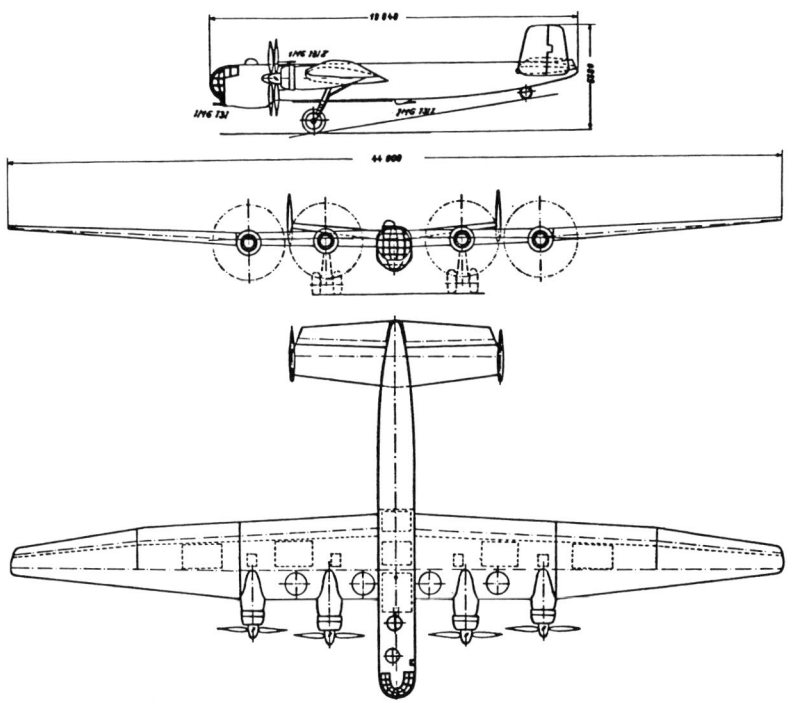

The He 274 Aircraft Type Sheet of 1 January 1942 showing a defensive armament configuration consisting of two MG 131Zs and one MG 131I machine-gun.

The He 274's planned lower nose gun barbette would have looked similar to this example, already tested on the Ju 288 mock-up.

The nose section of the He 274 V1, incorporating a pressurised cockpit, was based on that of the He 177A-4.

have only relatively light armour protection – the same as that projected for the He 177A-4 – but to save weight the engines and their radiators were to be left unprotected.

An even more telling weight-saving measure was the reduction of the bomb-load. This was requested by General Christiansen and the RLM representatives present at the two-day conference in January 1942, who considered a take-off weight of 29,500 kg (65,036 lb) to be the optimum for the required high-altitude performance. A request submitted by Heinkel in mid-January 1942 to lengthen and widen the cockpit was refused by the RLM, the basic guideline for the whole project being to avoid all time-consuming redesign and reconstruction, such as that involving the undercarriage and the cockpit, and to reduce such work to the minimum on other sub-assembly groups.

Construction of He 274 airframe parts was to be carried out by Farman personnel in an attempt to take some pressure off the Heinkel design bureau at Rostock-Marienehe. The first He 274 was to be completed by spring 1943, and the plan called for some of the six prototype and pre-production series aircraft to be put at the disposal of *Oberstleutnant* Theodor Rowehl's *Aufklärungsgruppe/ObdL* (a special long-range reconnaissance outfit attached to the OKL) for operational trials.

It was only on 2 February 1942 that the *Technische Amt* of the RLM issued binding Technical Guidelines for the He 274. According to these, it was to be a four-seat high-altitude bomber and reconnaissance aircraft equipped with a pressurised cockpit and remote-controlled defensive armament. The pilot was to be the commander of the aircraft, while the co-pilot would also act as bombardier and frontal gunner. The radio operator would look after the remote-controlled dorsal gun barbette; the ventral barbette would be operated by an air gunner. Each of the five MG 131 heavy machine-guns was to have 1,000 rounds of ammunition.

Later on, according to the RLM, there would most certainly be a change to heavier defensive armament. The double-barrel FDL 131Z dorsal and ventral gun positions would be replaced by FDL 151Z barbettes, while an FHL 151/1 turret had to be considered for the tail position.

The He 274's offensive armament was to comprise all current bomb calibres (with the exception of the SC 50), LMA and LMB aerial mines, as well as aerial torpedoes. Initially, besides the PC 1400X, it was also intended to carry the HS 293 and 294 radio-guided bombs. The proposed radio equipment was to consist of the FuG X, PeilG 6, FuBl 2, FuG 16, FuG 25, and the FuNG 101 and 102. Navigation and safety devices corresponded fully with the equipment carried by production series He 177s. Plans also

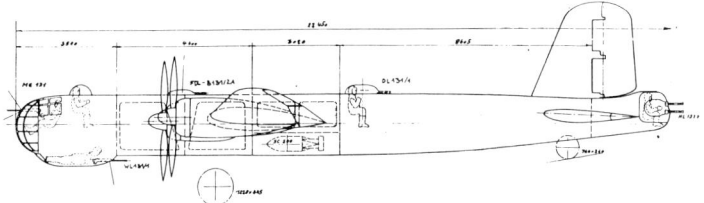
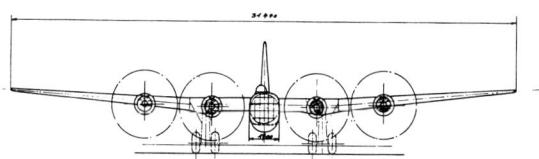
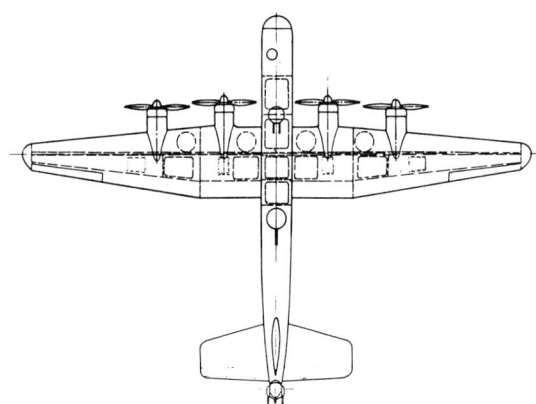

A three-view drawing of the He 274 armed with MG 151/20 cannon and a quadruple-barrel HL 131V tail gun turret.

called for Siemens K-12 navigation equipment to be fitted for automatic flight control.

Because of their intended high-altitude operations, all He 274s were to receive wing-, tailplane- and propeller-de-icing equipment, and each engine's exhaust system was to be fitted with suitable flame dampers. The wings were to be tightly riveted to permit their use as integral fuel tanks to house fuel necessary for the long climbing flight. After this fuel had been used, the wing tanks would be purged with carbon dioxide gas to render them explosion-proof.

As for the powerplants, Heinkel envisaged either the DB 605 with Argus engine-driven superchargers, the DB 624 or other high-altitude engines of similar power rating. According to the Heinkel performance department, four separate DB 605 engines would quite effortlessly lift the He 274 to altitudes of around 13,000 m (42,651 ft). The RLM's performance requirements for the He 274 included a speed of 480 km/h (298 mph) at low-level, increasing to about 650 km/h (404 mph) at 6,000 m (19,685 ft) altitude.

The 'Amerika Bomber'

As the Second World War escalated, so the OKL began to examine the possibility of carrying out harassment bombing raids against targets on the east coast of the USA, in particular New York. Such raids would have to be conducted

Belademöglichkeiten — He 277

Für Rumpfbreite 1500

Load	Abwurflast Kg	Kraftstoff Kg	Reichweite Km
2×SC1800, 3×SC500	5100	7700	3700
6×SC500	3000	9800	4750
2×SC2500	5000	7800	3700

Für Rumpfbreite 1750

Load	Abwurflast Kg	Kraftstoff Kg	Reichweite Km
2×SC1000, 2×SC1800	5600	11900	8600
6×SC500	3000	12200	11100
2×SC2500	5000	12000	9200

M.ehe 12.5.43.

A chart detailing the He 274's bomb-load options, ranging from six SC 500s to two SC 2500s.

by aircraft operating from and to airfields in Europe, but they would have to undertake the return legs of such missions without the aid of IFR. A request was issued by the RLM to the Germany's leading aircraft manufacturers for proposals for an aircraft to perform such raids, resulting in submissions from Messerschmitt (Me 264), Focke-Wulf (Fw 300) and Junkers (Ju 290). In addition, in May 1942, at the request of *Generalfeldmarschall* Milch, *Generalmajor Freiherr* von Gablenz examined the development status of all available or planned long-range aircraft that could possibly carry out such trans-Atlantic raids. The picture was not exactly bright.

Only two examples of the Me 261 were available, although the third airframe was also complete except for its two DB 610 powerplants. Designed and built in the late-1930s for record-breaking flights (including a planned non-stop flight between Berlin and Tokyo with the Olympic Flame onboard), the Me 261 was notable for its very deep, sealed wings which acted as integral fuel tanks for the DB 606A-1 (port) and B-1 (starboard) powerplants, each of which comprised a pair of coupled DB 601 12-cylinder liquid-cooled engines. The fuselage included rest bunks for the crew of five.

The Me 261 V1 (BJ+CP) first flew on 23 December 1940; the V2 (BJ+CQ) in spring 1941. The DB 610-powered V3 (BJ+CR) went on to conduct a 10-hour flight covering a distance of 4,500 km (2,796 miles) on 16 April 1943. The aircraft had a maximum range of 11,022 km (6,849 miles) at a maximum economic cruising speed of 400 km/h (248 mph) and subsequently served with *Oberstleutnant* Rowehl's *AufklGr/ObdL*, though its success in conducting long-range maritime reconnaissance tasks was hindered by a lack of defensive armament. In due course the Me 261s served with *E-Stelle* Rechlin's *Langstrecken Schule* (Long-Distance Training School) 'Wesenburg', and the Messerschmitt works, before the V1 and V2 were eventually scrapped.

Three examples of the Me 264, Messerschmitt's all-new and highly advanced response to the request for an ultra-long-range bomber to attack the USA were also under construction in mid-1942, of which the first, after work had recommenced on it in full measure, was expected to be completed in late 1942. Powered by four separate Jumo 211J-1 12-cylinder liquid-cooled engines with annular radiators (as per the Ju 88A-4), the Me 264 featured a circular-section

The unarmed Me 264 V1 (RE+EN), powered by four Jumo 211J-1 12-cylinder radial engines and first flown on 23 December 1942.

fuselage fronted by extensive nose glazing. Long-span, high aspect ratio wings housed 19,680 ltr (4,329 Imp gal) of fuel, while the tail assembly included inward-canted twin fin/rudders. No defensive armament was carried, but the bomb-bay in the centre-section of the fuselage could carry a 1,801 kg (3,970 lb) bomb-load.

By the time the Me 264 V1 (RE+EN) made its first flight over the Bavarian countryside on 23 December 1942, the USA had entered the war in Europe. As a consequence, it was realised that the Me 264 would require defensive armament and a greater bomb-load – requirements that increased the bomber's all-up weight and made it impossible to reach the USA and return home on the power of four Jumo 211J-1s. As *Generalmajor* von Gablenz noted:

> 'A military operation of any value cannot be achieved with the Me 264. Nevertheless, it is necessary to have a number of such aircraft in service, because during sporadic operations against the eastern coast of America, this aircraft in particular would be useful for armed reconnaissance flights over the Atlantic, as reconnaissance and contact aircraft for bomber formations, for co-operation with U-boats, dropping off radio buoys behind convoys so that they could be located by bombers and U-boats, and similar tasks.'

The Jumo 211J-1s were duly replaced on the V1 by four BMW 801G-2s, the aircraft going on to make several flights with these powerplants in early summer 1944. The new powerplants were also fitted to the V2 and V3, the latter intended as the prototype for the production series Me 264A.

The unarmed V2 had completed its taxying trials when it was destroyed by Allied bombers during an attack on Messerschmitt's Neu-Offing plant in late 1943. The V3, by contrast, was to feature the intended defensive armament suite of four MG 131s (one each in the nose, forward dorsal, and two beam positions) and two MG 151s (one each in the aft dorsal and aft ventral step positions). For the armed maritime reconnaissance role, three cameras and a 2,000 kg (4,409 lb) bomb-load were to be carried. The engines, crew positions and gun installations were to be armour-protected. Performance figures included a range of 14,996 km (9,318 miles) at a cruising speed of 350 km/h (217 mph). Given the increase in all-up weight, provision was made under the wings for the attachment of six rockets to boost take-off performance.

Unfortunately, due primarily to dwindling supplies of construction materials, but also because the estimated performance of both the Focke-Wulf Ta 400 and the He 277 was considered to be far superior, the RLM turned down the

A wind tunnel model of the improved Me 264 design.

The He 274 and He 277 Strategic Bombers

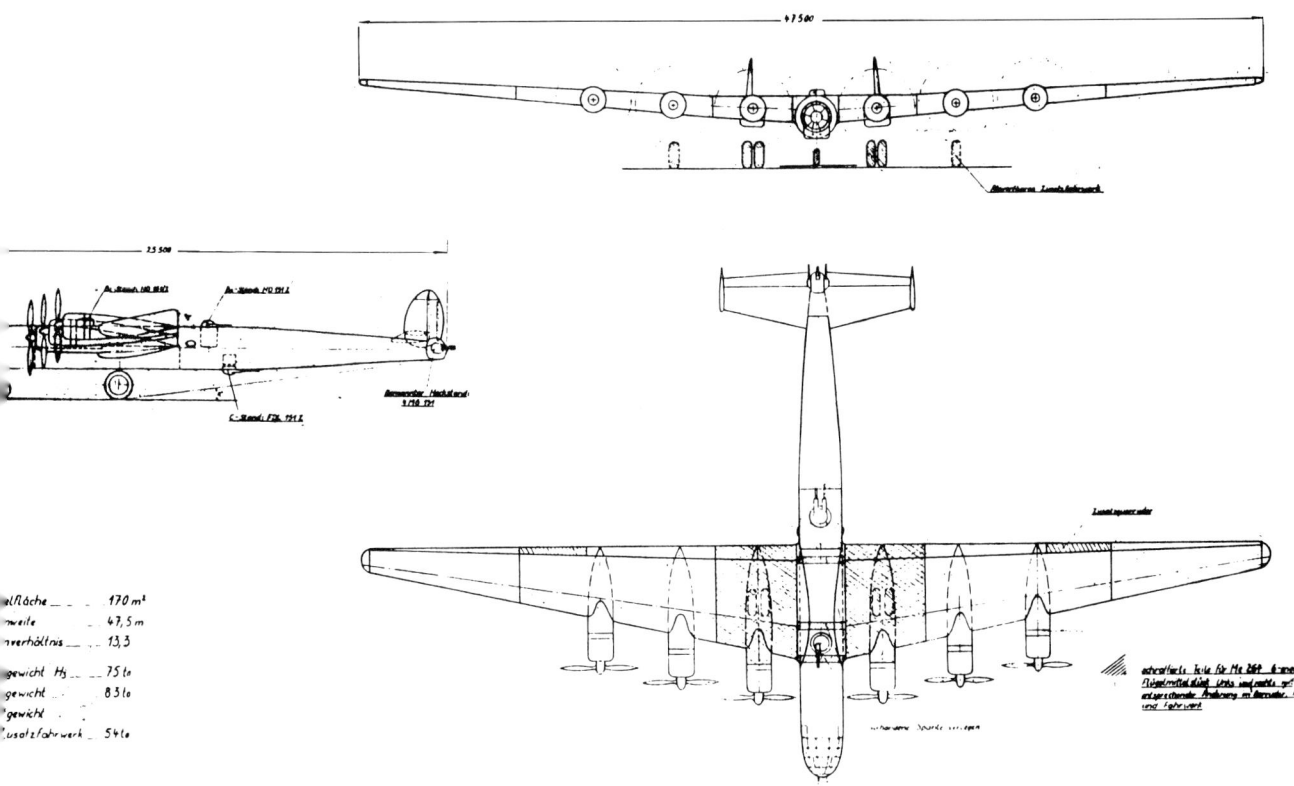

Three-view drawing of the heavily-armed Me 264B 'Amerika Bomber' powered by six engines.

Me 264 for series production in May 1943, before the V3 was completed. Just over a year later, on 10 June 1944, *Flieger-Ing* Böttcher conducted the final flight with the Me 264 V1, the aircraft subsequently being destroyed in an air raid.

As far back as late 1941, the Fw 300 long-range aircraft was in an advanced stage of development. At that time it seemed impossible to fly across the Atlantic Ocean and back without IFR support, which led in due course to Focke-Wulf submitting a specification for a new long-range bomber powered by four BMW 801Es with GM 1 (nitrous oxide) power boost units. This aircraft could lift a maximum bomb-load of 12,000 kg (26,455 lb) and had powerful defensive armament comprising twin-barrel MG 151Z cannon. In the structural description of 12 February 1943, the aircraft is estimated to have an all-up weight of some 53,500 kg (117,947 lb) and a range of 8,500 km (5,282 miles).

One month later, on 14 March 1943, Focke-Wulf's design office produced a description of a six-engined version of the Fw 300 with a range of 10,000 km (6,214 miles) and a 5,000 kg (11,023 lb) bomb-load. Powered by six DB 603s, each driving a four-bladed propeller, the aircraft was estimated to have an all-up weight of 80,000 kg (176,370 lb) and a maximum speed of 525 km/h (326 mph) at 6,000 m (19,685 ft) altitude. Later, the DB 603s were to be replaced by Jumo 213s or 222s, then under development. Defensive armament was to comprise one MG 151Z in the dorsal and ventral positions, plus one MG 151V-equipped turret; the total ammunition supply was to amount to 4,800 rounds (600 rounds per barrel).

In addition, there was the Focke-Wulf Ta 400 long-range multi-purpose aircraft, proposed in a company design document of 13 October 1943. To be powered by six BMW 801Es, the Ta 400 would carry 27,000 ltr (5,939 Imp gal) of fuel and have a range of over 9,000 km (5,592 miles). A bomb-load of 10,000 kg (22,046 lb) was envisaged, as well as Hs 293/294 radio-guided glide-bombs on

A wind tunnel model of the Ta 400, proposed by Focke-Wulf as a six-engined very long-range bomber.

external mountings. Defensive armament would comprise one FDL 103Z and two MK 103 cannon (nose), HDL 151Z barbettes (forward and aft dorsal), one FDL 151Z barbette (ventral) and one HL 131V (tail).

Although the Ta 400 was originally a response to the RLM's realisation that it would take six engines to power a long-range bomber equipped with defensive armament and a useful bomb-load to the east coast of the USA and back to Europe, Focke-Wulf saw the aircraft's primary role as long-range maritime reconnaissance/anti-shipping duties over the Atlantic Ocean, as well as to combat enemy aircraft operating against *Kriegsmarine* U-boats. Thanks to favouritism on the part of the RLM, the Ta 400 would later pose the most serious challenge to the He 274's chances of success.

The third manufacturer to respond to the RLM's 1941 request for proposals for a long-range trans-Atlantic bomber was Junkers, which offered the four-engined Ju 290, a development of the Ju 90B-1 commercial transport; itself, ironically, a development of an earlier Junkers long-range heavy bomber, the twin-engined Ju 89 'Ural Bomber'.

With the outbreak of hostilities in September 1939, most of the Ju 90B-1s were pressed into *Luftwaffe* service. The third B-1 for Lufthansa (D-ADFJ/Wk-Nr 90 0003) and the first of two Z-2s (B-1s powered by Pratt & Whitney SC3-G Twin Wasps) for SAA (ZS-ANG/Wk-Nr 90 0004) became the Ju 90 V7 (GF+GH) and V8 (DJ+YE) prototypes in support of the Ju 290 development programme; both sported a lengthened fuselage to improve handling. The V8 was fitted with full defensive armament comprising a forward-firing MG 151 and an aft-firing MG 131 in a ventral gondola beneath the nose and offset to port, an MG 151 (forward dorsal), another MG 151 (tail), and a single MG 131 on each side of the fuselage. This defensive armament configuration reflected the intention to use the Ju 290 for armed maritime reconnaissance tasks as well as transport duties.

The seventh Ju 90B-1 (D-AFHG/Wk-Nr 90 0007) was rebuilt to become the Ju 90 V11 and then the definitive Ju 290 V1 (BD+TX). First flown in August 1942, the V1 was distinguished by rectangular instead of circular windows in the fuselage, greater wing span and area, and more angular fin/rudder endplates. Unarmed (though the ventral gondola was retained), the aircraft was powered by four BMW 801MA 14-cylinder radials instead of the BMW 132H nine-cylinder radials used on the V7 and V8, development of the BMW 139 having been cancelled.

During a development discussion on 27 April 1943 it was ascertained that the Ju 290 powered by

The third Ju 290A-3 built, complete with nose-mounted FuG 200 Hohentwiel ASV radar aerial array for maritime reconnaissance tasks. Mission duration was up to 18 hours.

A close-up view of the Hohentwiel nose-mounted aerial array, as fitted on this Ju 290A-5 assigned to FAGr 5. A total of 11 A-5s were built.

four BMW 801D 14-cylinder radials and with an all-up weight of 41,000 kg (90,389 lb) would have a range of 6,600 km (4,101 miles). A heavier airframe with an all-up weight of 49,000 kg (108,026 lb), planned for production from summer 1944 onwards, was estimated to have a maximum range of 7,500 km (4,660 miles). Carrying an offensive load of 2,000 kg (4,409 lb), e.g. one radio-guided glide-bomb, it was calculated that the Ju 290 would still have a tactical range of 6,500 km (4,039 miles).

The three Ju 290A-9s (built before the A-7s and A-8s) featured increased fuel capacity which gave them a range of 8,785 km (5,459 miles); while the A-7 reconnaissance-bomber (the most-produced variant: 14 built), flight-testing of which started in spring 1944, had a 3,000 kg (6,614 lb) bomb-load that could include up to three Hs 293/294s or PC 1400Xs, a range of 5,800 km (3,604 miles), and a maximum speed of 438 km/h (272 mph) at 5,000 m (16,404 ft) altitude. As for the A-8 reconnaissance-bomber, defensive armament comprising no less than 10 MG 151s and one MG 131 earned it earned the title of most heavily-armed bomber of the Second World War.

In terms of offensive armament, no operations with glide-bombs were conducted by Ju 290As. The majority of the aircraft that reached operational units served with FAGr 5, formed on 1 July 1943 with Ju 290A-2/-3s. Using their FuG 200 *Hohentwiel* ASV radar, they would locate Allied convoys in the Atlantic Ocean, then pass on details of their location to marauding U-boats.

New Ju 290A-4/-5s were duly assigned to FAGr 5, but there were never enough aircraft available to make the unit's operations truly effective. Indeed, after the first 23 operations over the Atlantic the FAGr 5 command expressed their wish to immediately change over to the six-engined Ju 390 with its greater range, instead of waiting for more-powerful Ju 290 variants. Apart from that, the operational crews requested greater defensive armament consisting of twin-barrel MG 151Zs. In addition, a 30 mm MK 103 cannon for the tail position was seen as a real boost to combat value. In the event, FAGr 5 was withdrawn from France in August 1944, some of its aircraft going to I/KG 200 for clandestine transport operations.

Some months earlier, in November 1943,

The Ju 390 V1 (GH+UK) sported a lengthened fuselage and increased wing span, and was powered by six BMW 801D radial engines.

Junkers works pilots Dantzenberg and Matthias carried out IFR trials in support of the ultimate goal: bombing operations against targets on the east coast of the USA. The IFR trials (conducted using one Ju 290A-4 and one A-5) coincided with the start of work on the Ju 290B long-range heavy bomber with increased range and defensive armament. In the event, however, construction of the sole Ju 290B-1 prototype was cancelled on 30 June 1944, a victim of the decision to cancel the entire Ju 290 programme due to the increasing shortages of strategic materials.

As for the six-engined Ju 390, essentially a scaled-up Ju 290A devised early in the Ju 290A development phase to undertake long-range transport, maritime reconnaissance and bombing tasks; despite great effort Junkers managed to complete only two examples. The RLM, attracted by the potential for components designed and built for the Ju 290 to be used in the Ju 390, thus saving development time and expense, had in fact ordered three prototypes, one for each of the aforementioned roles, with particular interest being shown in the aircraft's potential as an 'Amerika Bomber' when it was offered by Junkers in competition with the Me 264 and Ta 400.

The Ju 390 V1 long-range transport (GH+UK) made its first flight at Merseburg on 20 October 1943. Powered by six BMW 801D 14-cylinder air-cooled radial engines, the largest conventional aircraft ever built in Germany featured four twin-wheel main undercarriage units which retracted into separate bays in each of the inner four engine nacelles. The aircraft was subsequently lost in a crash in 1945.

The Ju 390 V2, configured for maritime-reconnaissance tasks, featured a longer fuselage, FuG 200 *Hohentwiel* ASV radar, and defensive armament consisting of four MG 151s and three MG 131s. The aircraft was temporarily assigned to FAGr 5 in January 1944 for evaluation purposes, and went on to prove its capabilities by flying to within 19.3 km (12 miles) of the east coast of the USA, north of New York, and back.

Such operational range should have made the Ju 390 V3 long-range heavy bomber an awesome prospect. However, development and construction of the third prototype was accorded a low priority by the RLM, which instead placed greater emphasis on the Ju 290 and the Ju 390A-1 reconnaissance-bomber, the latter capable of carrying up to four glide-bombs on separate underwing ETCs and armed with three MG 151Zs, two MG 131Vs and two MG 151s. Carrying a 1,931 kg (4,257 lb) bomb-load, the Ju 390A-1 was estimated to have an operational range of 9,252 km (5,749 miles).

The Ju 390 V3 was never completed, and the Ju 390A-1 programme fell victim to the decision in June 1944 to halt all further development and production of the Ju 290/390 family. Events had overtaken the programme, and the scheduled start-up of Ju 390A-1 production in spring 1945 was now seen as being too late to enable such aircraft to carry out their intended tasks.

In the meantime, *Generalfeldmarschall* Milch had rejected a proposal from Blohm und Voss for a long-range land-based bomber version of the company's Bv 238 six-engined flying boat. Dubbed the Bv 250, the bomber would have sacrificed the Bv 238's planing bottom in favour of bomb-bays and a multi-wheel main undercarriage. Four prototypes were ordered, three of which were in pre-assembly when the Bv 238 programme was cancelled in summer 1944.

Following evaluation of the Messerschmitt and Junkers aircraft, *Generalmajor* von Gablenz turned his attention to the He 177/274 bombers:

'The Heinkel works has produced the He 177, which at present represents the aircraft with the greatest range. But even a newly-submitted project [the He 274] still does not have the definitive range.'

This, in effect, concluded the elimination process for a long-range high-altitude heavy bomber, although the He 274 was still in with a chance. But the deteriorating military situation and the lack of vital raw materials put tight restrictions on any planned series production. Indeed, between October 1942 and late 1944 production of the Ju 290A-series (the only one of the aforementioned types to enter service in any numbers, if at all) amounted to just 50 or so aircraft, the majority of which served in the maritime reconnaissance and reconnaissance-bomber roles.

He 274 versus He 277

An internal discussion at the Heinkel works had shown that the company was in favour of accelerated production of the He 274, with final

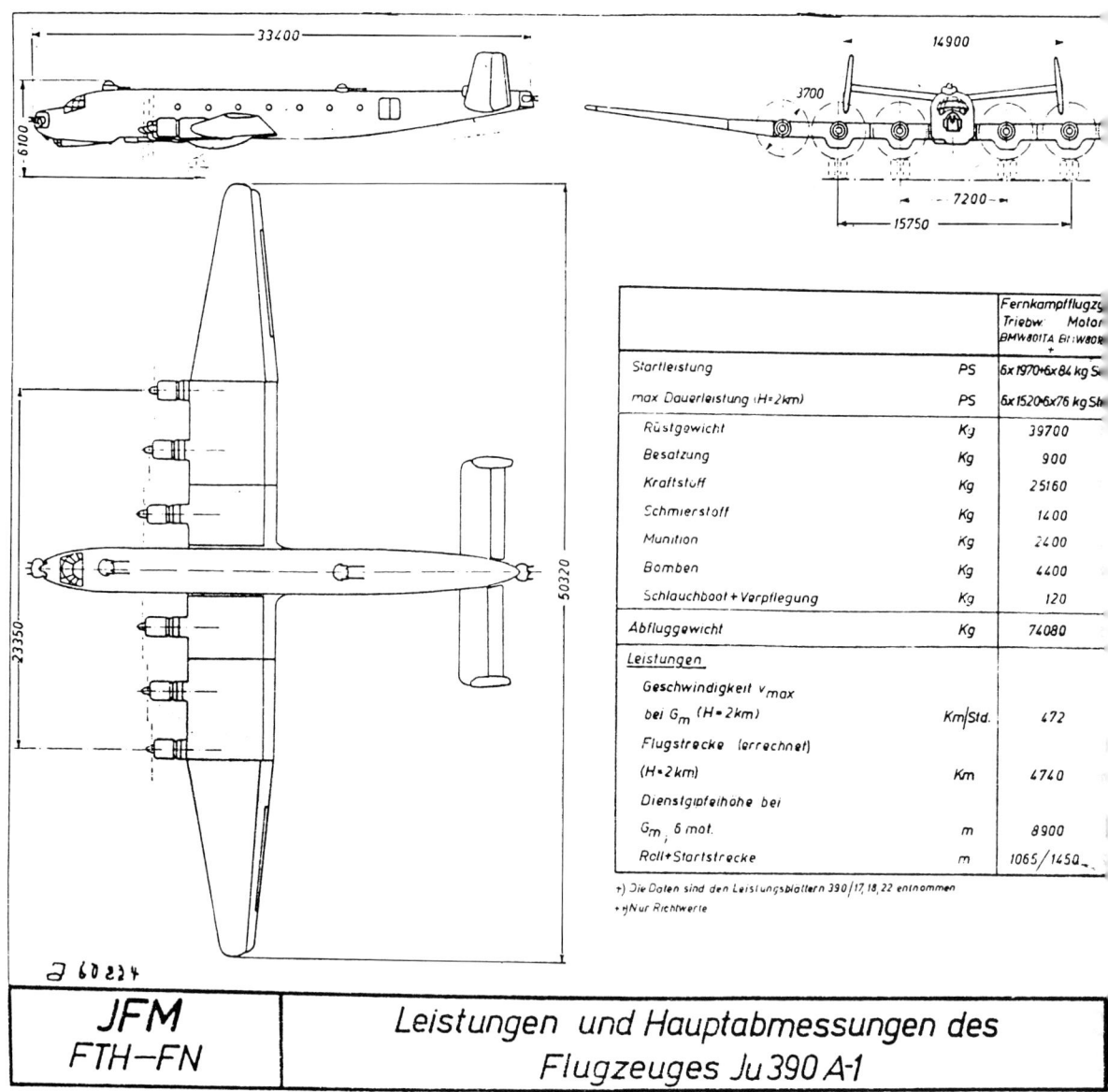

assembly of components manufactured in occupied France to be undertaken at the company's Rostock-Marienehe plant in an attempt to avoid further delays.

According to the Farman works at Suresnes, the first He 274 prototype was to be ready for delivery to Germany by 31 July 1943. However, at the same time Farman's Paris design office indicated that this deadline could not be met. Apart from the still unresolved question of which powerplant to install, there were numerous technical problems with the hydraulic system and the structural strength of the new, two-spar 44.22 m (145 ft 1 in) wing with detachable outer panels. Matters were made worse when, on 14 August 1942, having investigated further the plan to undertake final assembly in Germany instead of France, a stark conclusion was reached: if the Farman works actually delivered the first He 274 sub-assembly parts to Rostock-Marienehe in July 1943, the

A September 1944 three-view drawing of the planned Ju 390A-1 with quadruple-barrel nose and tail gun turrets.

aircraft could not be completed before early January 1944.

After the manufacture of several prototypes, construction of He 274A-0 pre-production series aircraft would commence, initially at Rostock-Marienehe. The first A-0 was expected in September 1944; altogether, it was planned to complete 27 pre-production aircraft by 31 May 1945. However, Farman reported further delays in their production run-up and indicated that delivery of sub-assembly parts probably could not be achieved within the prescribed time period. The reason given was the previously ordered re-arrangement of materials.

After an impassioned plea from Prof Dr-Ing Heinkel, Dipl-Ing Scherer of the Heinkel-Farman Liaison Bureau responded with the following binding deadlines to the RLM:

- January 1943: Delivery of He 274 V1 technical drawings
- August 1943: Completion of He 274 V1
- September 1943: Transport of He 274 V1 to Vienna-Schwechat
- January 1944: He 274 V1 cleared to fly

It was intended to complete the necessary wind tunnel tests by early spring 1944, provided the components needed for the ultimate breaking load tests were delivered by October 1943.

All further work on the He 274 was to take place at Rostock-Marienehe, with the emphasis on the new wings, longer fuselage and revised main undercarriage. At the same time the DVL would carry out research on the new outer wing panels, while Heinkel-Süd at Vienna-Schwechat would concentrate on the new twin fin/rudder assembly.

A further setback was feared in October 1942, when 47 Farman technical specialists were earmarked for transfer to Germany for other duties. Not only that; the powers-that-be in the *Luftwaffe* wanted to transfer an additional 80-90 men from the second Farman plant, and it was only when vigorous protests were lodged by Heinkel that the order was rescinded. Even so, on 12 October 1942 Prof Dr-Ing Heinkel expressed serious doubts about the chances of the work at Farman's Suresnes plant ever being completed:

'To me it looks altogether bleak for the He 274 in France, and I am ever more convinced that once Schwechat is fully established and running we should fit in the He 274 more there during the coming spring.'

Yet despite Heinkel's doubts and all the information to the contrary, *Oberst* Pasewaldt decided on 19 February 1943 that all work on the He 274 still being carried out in Germany should be continued in France, and that construction of the

A full-scale wind tunnel model of the He 274's twin fin/rudder tail assembly under test at Meudon.

Construction of the He 274 V1 and V2 prototypes was undertaken by Farman at the company's Suresnes plant outside Paris.

The He 274 and He 277 Strategic Bombers   195

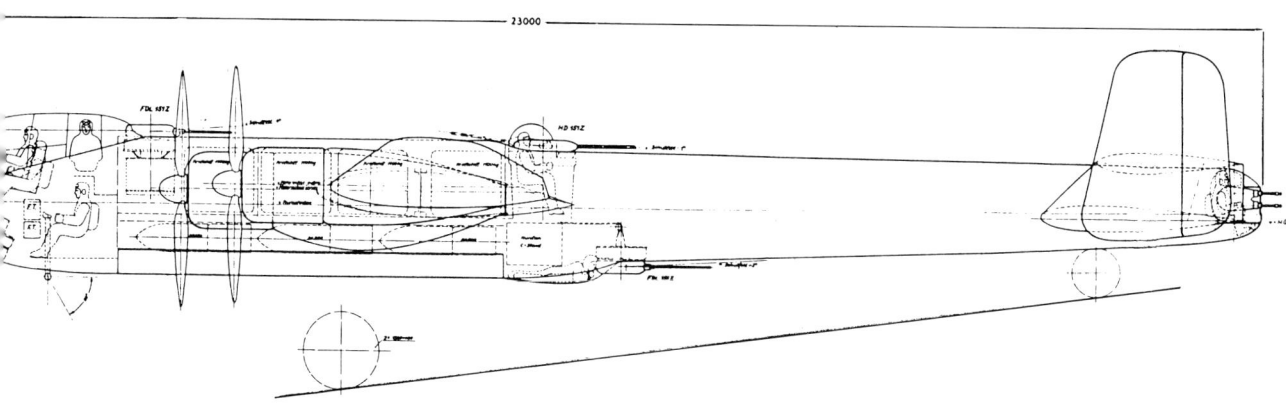

Side-view of the He 277 powered by four BMW 801 radial engines and with a crew of seven.

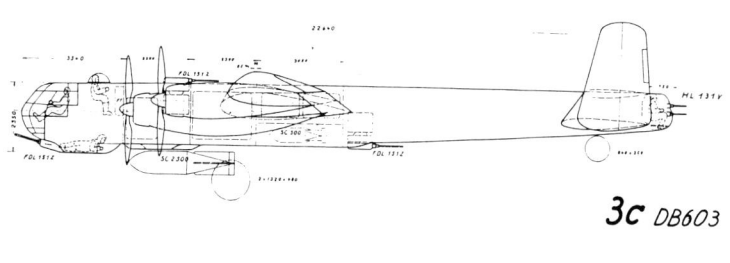

3c DB603

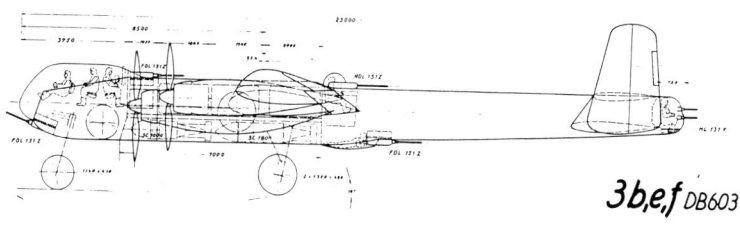

3b,e,f DB603

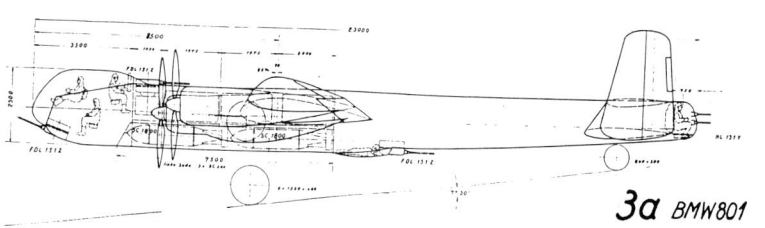

3a BMW801

Side-views of three planned versions of the He 274 powered by DB 603 or BMW 801 engines.

prototypes should begin at the Suresnes plant without delay!

In addition to the He 274, Heinkel also hoped to build the He 277 heavy bomber. Using an He 177A-7 fuselage, but with the ventral gun position moved further aft (as per the He 274) and an MG 151Z in the tail turret, the He 277 was to be powered from the outset by four separate engines, namely BMW 801E radials, carried on a wing increased in area to 130.00 m≈ (1,399.31 ft≈).

Powered by four BMW 801Es or Jumo 222A/Bs the He 274 was estimated to be a good 50 km/h (31 mph) faster than the He 177A-7. It could also safely be assumed that four separate engines would considerably reduce the fire hazard compared to the troublesome DB 606 and DB 610 coupled powerplants.

But the grand plans of the more powerful He 277 were already decreasing in importance. It was becoming increasingly obvious that the new cockpit could not be built on time, and it was not possible to increase the bomber's all-up weight to 38,000 kg (83,776 lb) without considerable strengthening of the undercarriage; failure to do so would mean the promised delivery deadline of summer 1945 could not be met. To complicate matters still further, it was forecast that by late 1945 all four-engined bombers in *Luftwaffe* service would need to be fitted with considerably more powerful engines if they were to offer effective operational range.

To save valuable development time, the Heinkel works envisaged fitting the so-called 'standard powerplants' on the He 274 from the outset. The first of these engines (known as 'powerpacks') had been flight-tested on a Do 217 early in 1943. The He 274 was also officially listed under Priority Rating DE, which afforded the project a considerable advantage in the allocation of more qualified workers.

By 3 July 1943 still no decision had been made concerning production of the He 277. The RLM seemed to favour heavy six-engined bombers at the expense of four-engined types – but only if a successful start was made to production of the Jumo 222 powerplant. According to the

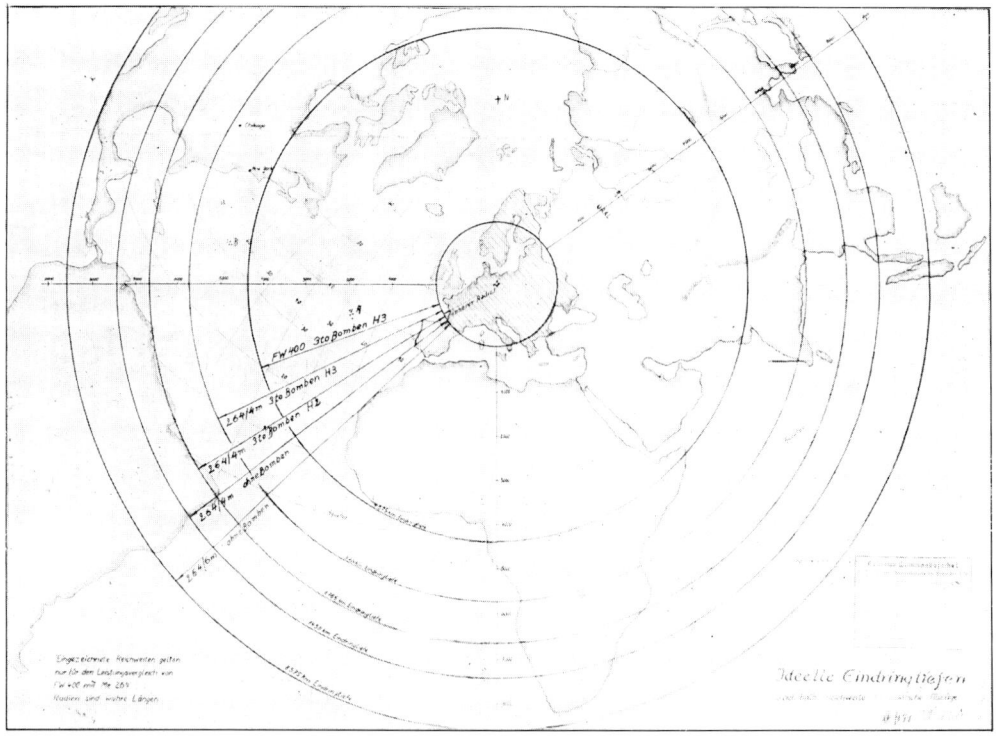

Calculated depth of penetration of the Ta 400 and the four- and six-engined versions of the Me 264 dated 12 May 1943.

prevailing opinion of the technocrats in Berlin, the four-engined He 277 had already been superseded as a viable replacement for the He 177 and should be substituted by the Ju 290 as quickly as possible. As far as the He 274 was concerned, it was only of interest if Heinkel could equip the aircraft's engines with TK 15 turbo-superchargers. To help the He 274 reach its optimum operational altitude, it was suggested that all defensive armament and armour protection be removed and fuel capacity reduced.

On 3 July Director Francke submitted the latest He 277 project data to *Fliegerstabs-Ing* Friebel. Having studied the data, the RLM representative envisaged the following as the most advantageous solution to the problem of finding a short-term replacement for the He 177:

> 'Building of He 277 with 133.00 m² (1,431.60 ft²) wing and four Jumo 222 engines. If Jumo 222 not yet available, flight tests with four BMW 801 up front. As an alternative solution in case Jumo 222 is a dud, six times BMW 801.'

The He 277's wings were to be constructed in such a way that it would be possible to attach two additional engines outboard of the existing four without the need for considerable alterations or redesign.

A heated discussion then took place as to whether the He 277, being only 50 km/h (31 mph) faster than the four-engined Ju 290 and six-engined Ju 390, really represented a worthwhile and acceptable advance in development; especially as the construction expenditure required for series production of the He 277 was by no means inconsiderable.

Director Francke relied mainly on the more favourable drag characteristics of his He 277 and pointed out that thanks to its nosewheel undercarriage configuration, the He 277 would be superior to its rival developments. According to Francke, '. . . the Ju 290 could be described as a steamer, while the He 277 by contrast was not only an elegant but also an agile aircraft.' Asked about the Ta 400, Francke declared that it seemed very questionable to him whether this machine could actually achieve the performance estimated by Focke-Wulf. On the other hand, it was also doubtful whether the He 277 could achieve the claimed range of 8,000 km (4,971 miles)!

Despite such doubts concerning powerplant and performance, a teleprinter message from the *Technische Amt* on 23 July 1943 demanded the pre-construction of the He 277 powered by four Jumo 222s. In addition, the necessary tests and calculations for a version fitted with a 170.00 m² (1,830.00 ft²) wing were to be undertaken as quickly as possible.

Less than a month later, on 20 August 1943, Director Francke reported that the authorities in Berlin were considering the deletion of the Ju 290 from production in favour of the He 277, the performance spectrum of the latter having found more approval.

Early in September 1943 the Heinkel works finally received the calculated performance data of the Ta 400, which made it possible for Heinkel to prepare a performance-and-weight comparison between Focke-Wulf's proposal and the He 277.

	He 277	Ta 400
Crew:	4	6
Powerplant:	6 x BMW 801E	6 x BMW 801E
Wing Span:	45.00 m	42.00 m
	(147 ft 7fi in)	(137 ft 9fi in)
Wing Area:	170.00 m²	170.00 m²
	(1,830.00 ft²)	(1,830.00 ft²)
Undercarriage:	Nosewheel	Nosewheel
Weight (empty):	31,730 kg	27,700 kg
	(69,953 lb)	(61,068 lb)
Weight (loaded):	60,000 kg	60,500 kg
	(132,277 lb)	(133,379 lb)
Useful Load:	28,270 kg	32,800 kg
	(62,325 lb)	(72,311 lb)
Maximum Speed:	560 km/h	575 km/h
	(348 mph)	(357 mph)
Range:	5,500 km	7,000 km
	(3,418 miles)	(4,350 miles)
Armament:		
Nose (lower)	1 x MG 151Z	2 x MK 103
Dorsal (fwd)	1 x MG 151Z	1 x MG 151Z
Dorsal (aft)	1 x MG 151Z	1 x MG 151Z
Ventral	1 x MG 151Z	1 x MG 151Z
Tail	1 x MG 131V	1 x MG 131V

The following month Prof Dr-Ing Heinkel drew up a wider performance comparison between the Me 264, Bv 222, Ju 290, Ta 400 and He 277. His work led to the following conclusions:

> 'In our opinion, the Me 264 is a record-breaking aircraft and does not come up to service requirements for operations in large numbers. The Bv 222

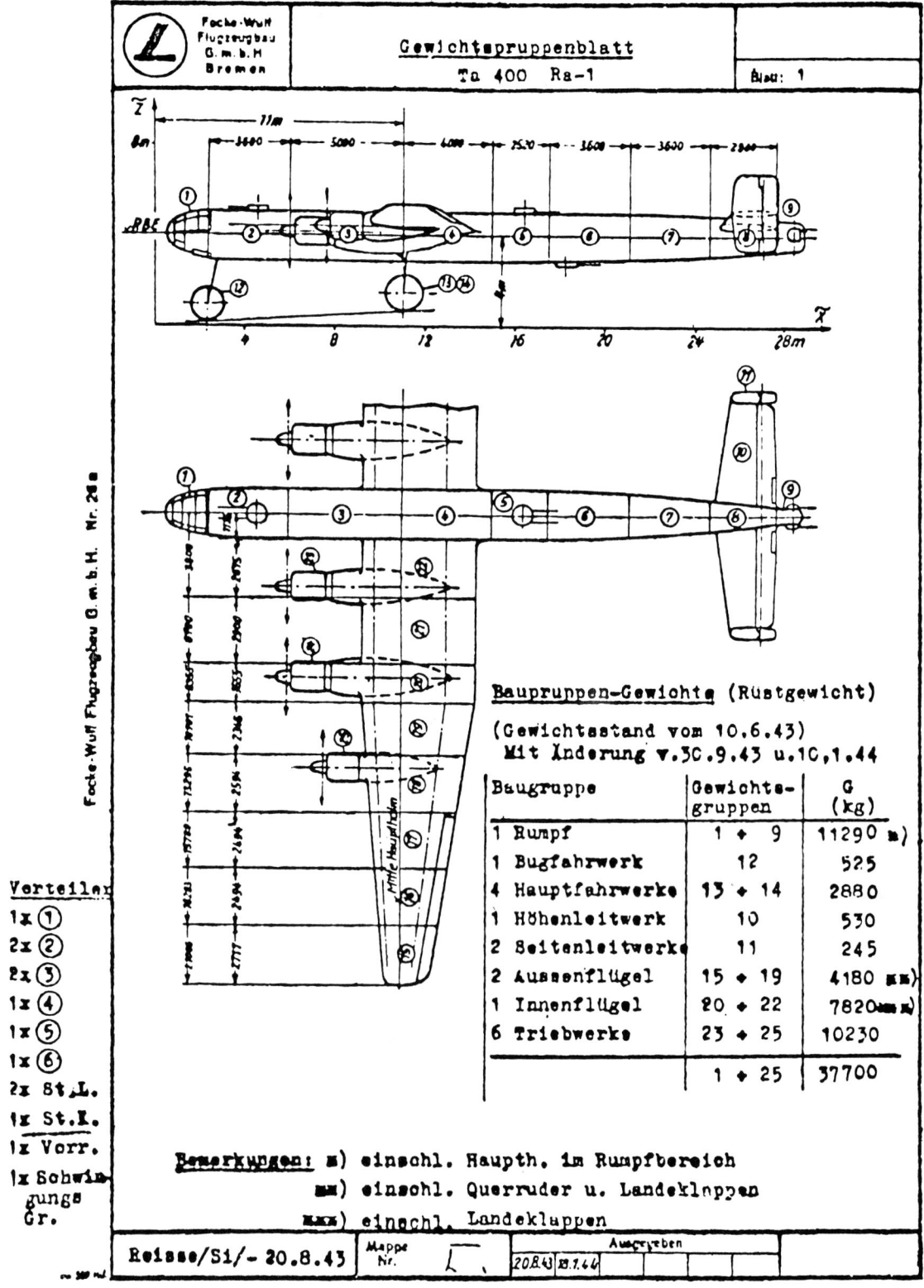

A weight grouping sheet for the projected Ta 400 Ra-1 dated 20 August 1943.

and Ju 290 are far too big and are not bombers, in addition to which the Ju 290 has to be altered to Ju 390 (six engines). This would make the construction effort bigger still. Thus, only the Ta 400 and the He 277 remain as useful operational aircraft.'

Just eight days later Focke-Wulf's challenge to the He 277 was brought to a close when the *Technische Amt* unceremoniously abandoned all further development of the Ta 400. In the meantime, however, early design work was underway on a new rival bomber that would come to pose a serious threat to the Heinkel bombers – the Ju 488.

Available Options?

During a meeting of RLM heads of departments on 7 December 1943, the subject of high-altitude bombers then available to the *Luftwaffe* came up for discussion. *Oberstleutnant* Knemeyer (GL C/E) first presented an overview of the development situation of the relevant aircraft.

Two of the aircraft under development, the Hs 130E-0 and Do 217P, were twin-engined designs powered by DB 603 12-cylinder liquid-cooled engines, and both were fitted with a fuselage-mounted DB 605T engine to provide power for a centrally-mounted two-stage *Höhenlader Zentrale* (HZ: built-in high-altitude compressor) which fed pre-compressed air to the wing-mounted engines in the thin atmosphere found at high altitudes.

Henschel Flugzeugwerke AG had played a leading role in advancing Germany's understanding of high-altitude flight and the development of pressurised cabins and engine superchargers, not least with its Hs 128 V1 and V2 research aircraft built in the late 1930s. These were subsequently developed as the Hs 130A, intended primarily for high-altitude reconnaissance tasks.

The Hs 130A V1 (GH+OM) took to the air on 23 May 1940, followed on 17 July 1940 by the V2 (GH+ON). Both aircraft carried a two-man crew in a circular-section fuselage fitted with a pressurised cockpit fronted by limited glazing. Power was provided by two DB 601Rs with a single-stage supercharger. These aircraft were followed by five Hs 130A-0s, used mainly to test various alternative powerplants.

The use of the Hs 130 as a high-altitude bomber was proposed by Henschel in early 1941 in the form of the Hs 130B, but this project was abandoned at the mock-up stage because of continuing problems with the intended DB 601D engine. Next came the Hs 130C, work on which began in late 1940. This model featured a four-man crew in a pressurised cabin with greatly increased glazing and was a late entry in the 'Bomber B' competition. Three Hs 130C-0s were built and flown between late 1942 and early 1943, but plans to manufacture an Hs 130C-1 production series

A wind tunnel model of the Henschel Hs 130C-0, three of which were built as prototypes of a high-altitude bomber. The first example (NK+EA/Wk-Nr 0011) flew on 10 November 1941.

bomber came to nought with the cancellation of the entire 'Bomber B' programme in summer 1943.

The Hs 130D, based on the Hs 130A but powered by DB 605s, met with a similar lack of success, the primary reason for its abandonment being problems with the intended two-stage supercharger. But Henschel persevered with the idea of supercharged power in the Hs 130E, its two DB 603 engines being augmented by a DB 605T mounted inside the fuselage to power the HZ. First flown in prototype form in July 1942, this three-man high-altitude bomber was to carry a 1,800 kg (3,968 lb) bomb-load on underwing racks.

Just as the Hs 130 drew heavily on the design and construction of the Hs 128, so the Do 217P made use of the fuselage, wing, undercarriage units and tail assembly of the Do 217E-2. The Do 217P also featured large intercooler radiators beneath the inner wing sections and prominent air scoops low on the fuselage just aft of the wing trailing-edge. A four-man crew was housed in an extensively-glazed pressurised cabin.

The Do 217P V1 (BK+IR/Wk-Nr 1229) conducted its maiden flight on 6 June 1942, and flight-testing with the HZ in operation began the following August. The aircraft went on to reach an altitude of 13,650 m (44,783 ft); but this was subsequently bettered by the Do 217P V4 which reached 15,200 m (49,869 ft).

In addition to the HZ-equipped Hs 130E and Do 217P there was also the Ju 186, a progressive development of the Ju 86P-1/R-2 (high-altitude bombers) and Ju 86P-2/R-1 (high-altitude reconnaissance). Although capable of reaching altitudes in excess of 16,356 m (47,000 ft), these long-span models based originally on the outdated Ju 86D bomber were increasingly vulnerable to interception by enemy fighters, most notably the Spitfire. Thus, ever higher altitudes were sought, one line of development being the Ju 186 powered by four Jumo 208s or two Jumo 218s (coupled Jumo 208s) and fitted with a three-seat pressurised cockpit.

In fact, calculations showed that the Ju 186's operational ceiling did not meet the requirements of the *Technische Amt* and, combined with problems with the development of the Jumo 208, the project was cancelled. Junkers was then asked to investigate whether a four-engined development of the twin-engined Ju 388 would be possible as a more efficient high-altitude bomber/reconnaissance aircraft. A short while later, however, this project was itself shelved when calculations revealed that such an aircraft would possess insufficient operational range.

Nevertheless, there remained the requirement

The first high-altitude bomber built in series, the Ju 86P-1 was primarily used for reconnaissance tasks (in conjunction with the P-2) and featured a two-seat pressurised cockpit. The aircraft was powered by two six-cylinder Jumo 207A-1s.

to produce a future standard model as a medium high-altitude bomber powered by Jumo 222s as quickly as possible, and to concentrate on the development of the two-stage compressor as well as the exhaust-driven turbo-supercharger. To this end the Ju 488 was seen by *Oberstleutnant* Knemeyer in a very positive light, and from 1945 he envisaged this aircraft covering the operational spectrum of the Ju 388 as well.

The question then arose, should development of the Hs 130E with its HZ installation be stopped soon, or temporarily continued with four engines to produce an efficient high-altitude reconnaissance aircraft? Although both the Hs 130E and Ju 488 were estimated to have a maximum speed of approximately 700 km/h (435 mph) at 13,500 m (44,291 ft) altitude, the Ju 488 was perceived to be more advantageous because its design incorporated many Ju 388 sub-assembly groups. During the meeting on 7 December 1943, *Oberstleutnant* Knemeyer noted the following points concerning development and production of the Junkers bomber:

> 'The Ju 488 has the long fuselage corresponding to that of the present Ju 88H. It will have the twin fin/rudder assembly of the Ju 288, the same tail gun position as the Ju 388 and will presumably be armed with a remote-controlled dorsal gun turret. It differs from the Ju 388 only by its wing centre-sections. The aircraft has four separate main undercarriage legs, and features the Ju 388 pressurised cabin.'

General Vorwald agreed with *Oberstleutnant* Knemeyer's choice, because he too considered development of the HZ-equipped Hs 130E to be far too slow. He also shared Knemeyer's view that the Ju 488 could carry a bomb-load of about 2,000 kg (4,409 lb) and have a range of 4,500 km (2,796 miles), thus making it useful as a maritime reconnaissance-bomber for operations out over the Atlantic Ocean. By contrast, the Hs 130E's range was only about 2,400 km (1,491 miles). Powered by four Jumo 222F engines using two-stage superchargers, it would have an estimated maximum speed of 710 km/h (441 mph), rising to 730 km/h (454 mph) at 15,000 m (49,213 ft) altitude.

As all those at the meeting agreed about the need to develop and produce the Ju 488, it was proposed to *Generalfeldmarschall* Milch (GLZ) that Junkers do all that was necessary to enable an immediate start to be made on this project. However, due to the limited material reserves, this decision in favour of the Ju 488 meant that it was not possible to simultaneously develop and produce the Hs 130E in quantity. Instead, orders remained temporarily at just 30 Hs 130E-0 pre-production and E-1 production series aircraft as a purely interim solution, until type manufacture came to an end.

In the event, no Hs 130E-1s were built before the project was finally abandoned during 1944. The same fate befell the Do 217P, of which four out of the six prototypes ordered had already flown. With the decision to produce the Ju 488 to meet the high-altitude bomber requirement, the Do 217P V5 and V6, both of which were all but complete, were scrapped in March 1944 and the development programme abandoned. Later that year, both the Do 217 V1 and V2 were destroyed in an enemy air raid. Development of the central turbocharger system was also halted during 1944.

As for the Ju 488, the RLM's enthusiasm was translated into an order for six prototypes (V401-V406). Aware of the problems that had plagued the He 177 from the outset, Junkers' solution to the pressing need for a reliable four-engined high-altitude bomber was to develop such an aircraft and cut its production time and costs by basing it heavily on sub-assemblies developed for the Ju 388K/L (forward fuselage, pressurised cockpit, outer wing sections), Ju 88A-15 (wooden ventral pannier), Ju 188E (fuselage mid-section) and Ju 288C (twin fin/rudder tail). The existing sub-assemblies would be combined with a new fuselage plug and wing mid-section (the latter to carry the two additional engines), both of which were to be manufactured at the former Latécoère works at Toulouse; power would be provided by four separate BMW 801TJs.

By mid-July 1944 the new fuselage and wing mid-sections for the Ju 488 V401 and V402 were ready for shipment by rail from Toulouse to the Junkers works at Bernburg where they were to be mated with the outer wing sections, tail assembly and powerplants. But on the night of 16-17 July the fuselage sections were sabotaged by members of the French Resistance, the damage being considered so severe that the sub-assemblies were duly written off.

Even before the loss of the new sub-assemblies at Toulouse, development of the Ju 488 V403 was already underway. This aircraft was to feature

General view of the BMW 801TJ radial engine with built-in supercharger.

some radical changes, most notably a wider, deeper and longer fuselage now minus the prominent ventral pannier. In addition, the entire wing was to be moved aft; defensive armament was to be added in the shape of two MG 151s in a remote-controlled aft dorsal barbette and two MG 131s in a remote-controlled tail barbette; and the BMW 801TJ 14-cylinder air-cooled radials were to be replaced by four Jumo 222A-3/B-3 24-cylinder liquid-cooled radials. The pressurised cockpit of the Ju 388K was retained. Fuel capacity rose from 5,682 ltr (1,250 Imp gal) to 15,065 ltr (3,314 Imp gal); the bomb-load from 2,000 kg (4,409 lb) to 5,000 kg (11,023 lb).

The effect of these changes on the estimated performance of the Ju 488 V403-V406 was dramatic. Operational range rose from 2,000 km (1,243 miles) to 3,400 km (2,113 miles) at an average speed of 490 km/h (304 mph) and 7,000 m (22,966 ft) altitude. Maximum speed was now 690 km/h (429 mph) at 7,200 m (23,622 ft) altitude.

And yet, despite such promising performance estimates, the Ju 488 was abandoned in late 1944 before any prototype was even completed let alone flown. In short, Germany's increasingly desperate situation as the Allies advanced through northern Europe effectively signalled the end of bomber development and the transfer of all relevant production capacity to the manufacture of fighters.

Earlier that year, in January, with development of the Ju 488 having been given the go-ahead, the RLM initiated a simultaneous investigation by Messerschmitt and Heinkel into the feasibility of mating an improved He 277 fuselage with the Me 264's wings. As the He 177 fuselage offered more potential in terms of accommodating greater bomb-loads and fuel reserves than that of the Me 264, the RLM actively considered a combination of both aircraft types. A strengthening of the defensive armament by fitting quadruple-barrel weapons in the nose and tail positions therefore corresponded fully with the RLM's wishes.

The optimistic estimates prepared by both Heinkel and Messerschmitt listed a take-off weight for the hybrid of 56,000 kg (123,459 lb), this including a bomb-load of 3,000 kg (6,614 lb) carried over a maximum range of 12,500 km (7,767 miles). To get airborne, however, such an aircraft would have had to have an auxiliary wheeled undercarriage, rocket boosters to assist take-off and a runway no less than 1,700 m (5,577 ft) in length.

The first serious calculations of the performance of such an aircraft were conducted in February 1944, but they indicated a range of only about 8,000 km (4,971 miles). There were also numerous detailed questions concerning the project that remained unanswered, and so E-Stelle Rechlin was immediately entrusted by the RLM with the task of finding solutions to as many of the outstanding problems as possible.

Apart from everything else, special emphasis was now put on strengthening the defensive armament of all long-range bombers. According to a discussion involving the RLM's departmental chiefs on 29 February 1944, the highest priority was to be given to fitting the Ju 290 with a powerful tail turret. But the planned production run-up of the HL 151Z, scheduled for mid-1945, was considered to be too late. Even the projected HL 131Z turret, offering as it did a better effective range of about 1,000 m (3,280 ft), did not seem to meet the desired operational requirements. Indeed, when intercepted and fired upon by RAF de Havilland Mosquitoes, the bomber crews had to be prepared for hits from the night-fighters' 20 mm Hispano cannon at a range of up to 2,000 m (6,562 ft).

As the concentration of fire of the quadruple-barrel turret and the greater ammunition capacity favoured the tail position, an order was placed for the first model of this turret. A pattern installation was to be ready for practical tests by the beginning of March 1944, and trials of both tail turrets – also of interest for application on the He 277 itself – were to follow at *E-Stelle* Tarnewitz as soon as possible. Another turret envisaged for inspection on 3 March 1944 was the first twin-barrel MK 103Z cannon installation, which was also intended for heavy bombers and reconnaissance aircraft.

While these discussions and evaluations continued apace, work on the He 274 V1 at the Suresnes plant outside Paris was going according to plan. The rebuilding of the selected He 177 airframe was beginning to take shape, and the Farman engineers had got as far as fitting it with the first all-wood twin fin/rudder assembly.

Suddenly, on 20 April 1944, three of the six He 274 prototypes as well as the option for the planned He 274A-0 pre-production series were cancelled. Only the He 274 V1-V3 and a static test airframe were now authorised to be built. Another victim of the red pencil was the safety cockpit with ejection seats. Shortly before this decision was announced, Heinkel was instructed by the RLM to suspend all work on the He 277. All components and parts in the process of manufacture were to be scrapped.

This complete renunciation of the recently and repeatedly requested four-engined bomber was also obvious during a conference between *Generalfeldmarschall* Milch and the *Jägerstab* in Berlin on 12 May 1944. Other participants included *Generalmajor* Vorwald, *Oberst* Diesing, *Generalstabs-Ing* Lucht and numerous representatives from Germany's aircraft manufacturers. By then, the reality of the limited material resources had finally caught up with the narrow-minded planners in the RLM. Even so, a certain amount of wishful thinking was still in evidence.

Thus, Director Thiedemann of JFM spoke optimistically about the impending production of the Ju 287 high-speed bomber, the first prototype of which was then under construction. Powered by four Jumo 004B-1 axial-flow turbojets (one under each forward-swept wing; one either side of the forward fuselage), this unusual aircraft's components included a modified He 177A-3 fuselage, the fin and rudder of a Ju 188G-2 and the wheels of a captured USAAF B-24 Liberator bomber. The type specification included a bomb-load of 3,000 kg (6,614 lb) and a range of 2,000 km (1,243 miles), the former rising to 4,000 kg (8,818 lb) and the latter to 1,900 km (1,181 miles) if rocket assistance was available during take-off. Maximum speed was calculated as 600 km/h (373 mph).

The Ju 287 V1 (RS+SA) first flew on 8 August 1944, each of its four turbojets being augmented by Walter 501 rocket packs that were jettisoned after take-off and parachuted to the ground for recovery and re-use. By then, however, all further bomber development and production had been officially abandoned, and so Director Thiedemann's hopes that the first 100 production series examples (powered by six BMW 003A-1s) would be completed by August 1945, with an additional 100 per month leaving the production line by December 1945 (provided Ju 88 night-fighter production could be reduced accordingly), were never realised.

War Booty

On 6 June 1944 – D-Day – Allied forces landed on the beaches of Normandy, their arrival serving notice that from then on the days of German occupation of France were numbered. At Farman's Suresnes plant, work on the He 274 V1 continued, and by early July 1944 the four-engined aircraft was almost ready for its first flight. However, as it could not be cleared for the transfer flight to Germany, and with completion of the He 274 V2 still some way off, drastic action was deemed necessary to stop the aircraft falling into the hands of the advancing Allied forces. With the airfield about to be captured, explosive charges were attached to the engines of both aircraft and

The He 274 V1 under construction at Farman's Suresnes plant.

An attempt to destroy the He 274 V1 before advancing Allied troops overran the Suresnes plant in summer 1944 failed and the aircraft was subsequently repaired.

detonated. But the attempt at sabotage failed; the airframes survived undamaged.

After Liberation, French technicians of the by now nationalised and renamed *Ateliers Aéronautiques des Suresnes* (AAS) fitted replacement engines to the He 274 V1 and V2, and the aircraft were redesignated the AAS 01A and 01B respectively.

Piloted by Commandant Housset along with flight engineers Gilbert Seimpere and Jean Maueon, the re-engined AAS 01A took to the skies on a successful first flight at Orléans-Bricy on 27 December 1945. Later on, during testing with the *Centre d'Essais en Vol* (France's equivalent of Great Britain's Royal Aircraft Establishment) at Bretigny, the 36,000 kg

Detailed view of the AAS 01A's nose section with its spacious pressurised cockpit. The aircraft was powered by four DB 603s.

Above and pages 206 and 207:
A series of views of the AAS 01A shortly before its first flight at Orléans-Bricy on 27 December 1945.

The He 274 and He 277 Strategic Bombers   207

(79,366 lb) aircraft reached altitudes of up to 15,000 m (49,213 ft). At an altitude of 11,000 m (36,089 ft) the AAS 01A registered a speed of approximately 600 km/h (373 mph).

Thanks to its high operational ceiling and a flight endurance of up to eight hours, the AAS 01A proved itself an ideal high-altitude research aircraft for the testing of pressurised cockpits. After the completion of this work the aircraft was used mainly as a platform for in-flight testing of prototype combat aircraft fitted with pulse-jet units and rocket motors. In 1948, the AAS 01A made several flights carrying a one-fifth scale flying model of the Sud-Ouest SO 4000. Later, the AAS 01A was used for practical trials of the Leduc 010 and 016 experimental ramjet-powered interceptors.

The AAS 01B too was powered by four DB 603A engines found amongst the spares left behind in France by the retreating Germans. The engines were fitted with TK 11 turbo-superchargers. The AAS 01B first flew almost exactly two years after the AAS 01A, on 27 December 1947, by which time AAS had been absorbed as part of the *Société Nationale des Constructions Aéronautiques du Sud-Ouest*. Both aircraft continued flying until the early 1950s, but were eventually grounded and scrapped at Istres in 1953.

The same fate applied to He 177s captured by Allied forces as they advanced through northern Europe and into Germany itself. The official report Disposal of Enemy Aircraft and Gliders (7 July 1945) lists a total of 55 He 177s found in Germany, all of which were either destroyed or damaged. However, before then several examples of the bomber had been captured intact at their bases in France, and at least two were subsequently restored to flying condition.

The He 274 V1 during ground tests and in the air. The aircraft flew for the first time on 27 December 1945, by which time it had been redesignated the AAS 01A.

Left and opposite: The AAS 01A was put to good use in the post-war years as a test platform for various new French aircraft. These photographs were taken during testing of a scale model of the Sud-Ouest SO 4000 in 1948.

210 Heinkel He 177-277-274

One of the few captured serviceable He 177A-5s, this example was equipped with the FuG 200 Hohentwiel *ASV radar and a 'special weapon' underfuselage bomb rack.*

This He 177A-5 of 2./KG 100 fell into Allied hands having suffered fairly extensive damage.

THE HE 274 AND HE 277 STRATEGIC BOMBERS   211

Two British soldiers in the wrecked cockpit of an He 177A-5 found on an airfield in northern Germany.

The first of these was an He 177A-5, an ex-II/KG 40 aircraft fitted with the elongated wings of the A-7 model and armed with a 20 mm MG 151/20 cannon in the upper nose gun position. Captured by members of the French Resistance at Toulouse-Blagnac in September 1944, and initially flown by the *Armée de l'Air*, the aircraft was subsequently passed to the Americans, fitted with American radio equipment at Villacoublay and marked with USAAF national insignia; then flown to the Royal Aircraft Establishment (RAE) at Farnborough in Hampshire by Squadron Leader Randrup of the RAF on 14 January 1945. Five days later the aircraft took off again, this time for the short flight to RAF Bovingdon in Hertfordshire, home of the recently-formed US Air Transport Service (Europe), for delivery to the USAAF

Above and overleaf (top):
Little more than the carcass of the He 177A-5/V38 (KM+TB/Wk-Nr 550002) remained when it was 'captured' at Prague-Rusiye in May 1945.

A view into the bomb-bay of KM+TB.

The ultimate destination of Wk-Nr 550256 was the USA itself, and on 9 February 1945 the aircraft set off on its delivery flight. Unfortunately, what was intended to be a short stay at the first stopover point, Paris-Orly, lasted nearly three weeks because of problems that required the replacement of one of the coupled powerplants.

The bad luck continued on 28 February, when a tyre-burst resulted in the hasty abandonment of the take-off run at the start of the recommenced delivery flight. As a result of damage sustained during the subsequent ground-loop, including a broken fuselage, the aircraft was written off on site.

The He 274 and He 277 Strategic Bombers

This He 177A-5 (GP+RY/Wk-Nr 550256) was modified to become one of the few He 177A-7 pattern aircraft in summer 1944.

The loss of the USAAF's first airworthy He 177 was a blow, but a replacement was soon at hand in the form of another airworthy He 177A-5, also previously assigned to II/KG 40 and captured by the French Resistance at Toulouse-Blagnac in September 1944. Like its predecessor, this aircraft had its *Luftwaffe* markings replaced by *Armée de l'Air* insignia and black-and-white wing and fuselage identity striping, and had the inscription 'Prise de Guerre' (Prize of War) applied in small letters on either side of the fuselage.

Whereas the USAAF's first He 177A-5 stayed in

Above and overleaf (top):
This ex-KG 40 He 177A-5 (F8+AP/Wk-Nr 550062) was captured by the French Resistance, hence the 'Prise de Guerre' inscription ahead of the fuselage roundel. The aircraft was photographed after being flown to RAE Farnborough, but before its Armée de l'Air *markings were replaced by RAF markings.*

Another ex-KG 40 found in France was this He 177 A-0 (DR+IU/Wk-Nr 32013), seen here after the application of **Armée de l'Air** *markings. The aircraft was subsequently scrapped.*

France until January 1945, this aircraft was ferried to the Experimental Flying Department (EFD) at RAE Farnborough on 10 September 1944 by Wing Commander Falk and Squadron Leader Pearse. The *Armée de l'Air* insignia were soon replaced by RAF markings, the *'Prise de Guerre'* inscription gave way to a large encircled yellow 'P' (for Prototype), and the serial TS439 was applied to either side of the rear fuselage.

On 20 February 1945, after examination and evaluation by the EFD, TS439 made the short flight from RAE Farnborough to the Aircraft and Armament Experimental Establishment at Boscombe Down in Wiltshire. This proved to be TS439's last flight; two months later, following the demise of Wk-Nr 550256 at Paris-Orly, TS439 was transferred to the USAAF, dismantled and shipped to the Air Technical Service Command's

Some remains of He 177s produced at Heidfeldt, such as this piece of cockpit framing, can still be found on the former factory site.

Foreign Aircraft Evaluation Center at Freeman Field, Indiana. Allocated the serial FE-2100 (FE: Foreign Equipment) once it had arrived in the USA, the aircraft was subsequently reserialled T2-2100 when it passed into USAAF storage at Oak Ridge, Illinois as a potential future museum exhibit. Sadly, like other ex-*Luftwaffe* aircraft held at the site, T2-2100 would appear to have been scrapped at some point during the second half of the 1940s.

Another captured He 177, this time a pre-production A-0 (DR+IU/Wk-Nr 32013) used as a training aircraft by IV/KG 40, remained in France for some time after the war in Europe had ended. Eventually, however, this aircraft, like so much other ex-*Luftwaffe* war booty, was broken up and reduced to mere scrap metal and raw material.

Apart from some He 177 remains found during construction work on the airfield at Zwölfaxing, as well as some He 177s that crashed into or made forced landings on lakes, not a single complete example of this German heavy bomber has survived for posterity.

The Balance

In 1938, Ernst Udet surprised his friend Ernst Heinkel with a preview of Germany's future air armament policy:

> 'In future, there will be no multi-engined bombers that cannot attack in a dive. The He 111 is the last horizontal bomber. Thanks to its accuracy on target, a medium-sized, twin-engined aircraft that delivers its 1,000 kg bomb-load on target in a dive has the same effect as a four-engined giant bird that carries 3,000–4,000 kg of bombs in horizontal flight but can only drop them inaccurately.
>
> We do not need the expensive machines that gobble up so much more material than a twin-engined dive-bomber. Junkers has the first twin-engined dive-bomber ready, the Ju 88. We can build two or three of this type with the same amount of material required for a four-engined job, and still achieve the same bombing effect. Jeschonnek too is fully enthusiastic. By building these large-size Stukas which cost less in materials, we can produce the number of bombers demanded by the Führer!'

As a consequence of this philosophy concerning bomber development, much of the development work already put in at the Heinkel works now seemed wasted. Although a little later it became evident that the twin-engined 'large-size *Stukas*' were not capable of any large-scale longer-range operations, development and production of the Ju 88 was still awarded the highest priority, and valuable time was lost.

Despite the repeated and strongly-worded objections of Ernst Heinkel that a big aircraft was unsuitable for diving attacks, his words went unheeded. The He 177, developed by Heinrich Hertel, was accordingly completely in line with the RLM's stated specifications:

> 'The ability of the He 177 to attack in a dive depends on the use of two powerplants. A normal four-engined machine cannot be made to dive, and is therefore eliminated on principle.'

For this reason, the He 177 was designed to be as small as realistically possible. At the same time, Hertel deliberately set the aircraft's all-up weight and dimensions on the small side. But the lack of sufficiently powerful and reliable powerplants, as well as the stated desire for a diving capability resulted in the use of what was, initially, a very troublesome coupled Daimler-Benz powerplant, the unreliability and poor maintenance of which could hardly have been worse.

After the He 177 had received a new cockpit, its behaviour in flight was dramatically improved as a result of fitting a new tailplane. It was also acknowledged that a longer fuselage would considerably improve the bomber's stability. Indeed, everything possible seemed to have been done to create an extremely efficient long-range bomber.

However, after factory and service trials it became evident that the He 177's wings were not strong enough for diving attacks. In addition to several expensive and involved airframe-strengthening processes, it became necessary to take into consideration hundreds of major and minor improvements before the impending series production could begin. Flying accidents and crashes then reduced the already small number of continuously airworthy prototypes, which inevitably hindered type-testing.

The troublesome coupled Daimler-Benz powerplants remained a constant source of trouble. It was only after comprehensive and very time-consuming development work that Daimler-Benz succeeded, to some extent, in getting the DB 606 units ready for operational service. Unfortunately it took far too long before the same could be said of the DB 610 as well. The unavoidable consequences of these problems were new and ever more delays. Yet, after rectifying faults in the cooling and air circulation systems, it seemed that the He 177 was finally ready for series production and deliveries of the first aircraft to the *Luftwaffe* could begin.

A pre-condition of the first use of the He 177 in anger was, naturally, that the crews would follow the operating instructions to the letter. Experience had shown that long-range flights with continuous power-loading were generally accomplished without any problems, but that any overloading of the powerplants could lead to damaged engines or engine fires. But that was not all.

Insufficient time was given over to preparing *Luftwaffe* bomber units for conversion onto the He 177; a problem that affected organisation and infrastructure on the ground, including personnel and technical support, as much as it did the training of aircrews. There were also instances when He 177s were delivered to units without having first received sufficient flight-testing. There was also a lack of suitable hangar space and parking facilities for large aircraft on operational bases thought to be outside the range of enemy bombers. In addition, there was a lack of good, well-trained workshop personnel and technicians able to maintain and service the new bomber.

In some cases, the complete re-equipment of individual bomber *Gruppen* within the prescribed time period failed due to low training levels and the lack of instruction for aircrews assigned to the He 177. Apart from that, for a long time there was a dire shortage of maintenance and servicing tools and equipment, not to mention replacement powerplants.

In May 1944 Major Schubert of the *Luftwaffengeneralstab* and *Reichsmarschall* Göring's Adjutancy was finally appointed to establish the principal reasons for the delays experienced in re-equipping *Luftwaffe* bomber units with the He 177. Nothing needs to be added to his report:

Most of the aircrew of units selected for re-equipment with the He 177 were operationally 'tired-out' and relatively few were from front-line units. The necessary personnel consisted primarily of young, often inexperienced aircrews, and for reasons of capacity their conversion training at operational training and replacement *Gruppen* could only be completed in relatively few cases. Most of the young pilots had only nine to 12 months of practical flying experience prior to being transferred to such a complicated aircraft as the He 177.

Apart from that, the new operational crews had been trained on the Ju 88, and most had hardly any training in the art of night-flying. The necessary conversion training meant the compulsory withdrawal of operational He 177s for use as trainers, which in turn led to an overload of work for the technical personnel due to the numerous instances of damage suffered by these aircraft as a result of the training activities.

Matters were made all the more difficult by the fact that some of the ground personnel had not been pre-instructed on the He 177. In addition, the vast majority of the technical personnel arrived at their He 177-equipped bomber *Gruppen* several months after the units had first received their re-equipment orders. By spring 1944, some units were still short of about 50 per cent of engine fitters. Some of the other personnel first set eyes on the He 177 upon arrival at their assigned unit's airfield, their instruction and training on the Heinkel bomber having to start there and then.

The supply of aircraft servicing tools and appliances also did not keep up with deliveries of He 177s. Thus, for instance, the wing attachment cranes needed to facilitate powerplant changes arrived several months after the delivery of the aircraft themselves, and even then they were too few in number. For IV/KG 1 there was no specialised engine-changing equipment at all, and for this reason the unit had to suspend all training activities in mid-April 1944.

The 'engine circulation' (service units – repair depots – service units) also did not flow as it should have done at first, because of a lack of transportation. Neither the supply of new engines nor the return of DB 606/610s in need of repair functioned properly, least of all the supply of exchange powerplants to individual airfields. It wasn't until April 1944 that these shortcomings were effectively overcome, but they were never fully eradicated.

According to Major Schubert, the time expenditure required for the maintenance and servicing of the He 177 was incomparable with that of any other operational aircraft in service with the *Luftwaffe*. The jacking-up operation to change the main undercarriage tyres alone (which had to done at least twice as frequently as on other aircraft types) lasted some 2fi hours using the prescribed mechanical spindle blocks. Yet by early summer 1944 far too few of these 12-ton spindle blocks recommended by the manufacturer were available to He 177-equipped units.

The layout of the powerplants too did not exactly help attempts to carry out the necessary servicing work. Because of the inaccessibility of the coupled engines their dismounting took considerably longer than similar work on, for example, the Ju 88 or He 111. Due to the low training level of the technicians, a 25-hour control check on the He 177 usually took two, sometimes even three days.

Criticism was also made of the airfields selected to receive the He 177. Apart from Aalborg in Denmark, all of the others were already completely overcrowded, and lacked the potential for dispersal, camouflage and suitable protection of their aircraft against bomb splinters and shrapnel. For this reason low-level attacks by Allied aircraft caused great losses amongst the He 177s parked out in the open from 1944 onwards, especially as the airfields were now constantly within the range of both fighters and bombers. To make matters worse, this vulnerability to attack had a knock-on effect on He 177 training activities, which sometimes had to be reduced by up to 50 per cent because enemy aircraft were on their way and air raid warnings came into force.

No consideration had been given to the fact that the technically complex He 177 required sufficient hangar space for maintenance and repair purposes, especially during the winter months. The delays caused by this shortcoming alone may well have been responsible for the postponement of He 177 operations by some six months to a year.

After the initial operations by I/FKG 50, the He 177 force never exceeded three incompletely-equipped *Kampfgeschwader*: KG 1, KG 40 and KG 100. This, and the type's inevitably late operational debut, as well as the increasing numerical superiority of Allied air power, prevented any large-scale He 177 operations in the West from 1943 onwards. The losses suffered in attacks on Atlantic convoys as well during the defensive operations against Allied landings, increased steadily and were soon so high that the *Luftwaffe* command had to suspend all further He 177 attacks.

The increasingly critical fuel shortage and the unavoidable decision of the defence authorities to allocate the highest priority to fighter production undoubtedly also led to the termination of He 177 production. Later, bombers such as the He 274 were granted at most a little extra time for completion.

The claim that simply because of the numerous engine fires the He 177 had become a '*Reichsfeuerzeug*' (State Lighter) cannot really be upheld, for not all of the technical defects and difficulties described here were caused exclusively by the coupled Daimler-Benz powerplants, although in many cases there was a strong connection. As so often happens, a multitude of minor causes can have a big cumulative effect that not infrequently results in the sad loss of an aircraft and its aircrew.

In truth, the blinkered technocrats of the RLM, without any feeling for, or real understanding of, the protracted effort involved in the development of a modern bomber, carried as much responsibility as some First World War fighter pilots who had gained rank and prominence after 1933, but were completely overwhelmed by modern technology on numerous occasions.

Last but not least, the main portion of blame should probably be ascribed to the advocates of an unrealistic doctrine of air warfare, and who were astutely appraised and judged quite early on by the famous Heinkel designer Siegfried Günter:

'They really are somewhat crazy with their dive-bombing. It has already become like a mania!'

Appendix 1

He 177 Prototypes

	V1/V2	V3	V4	V5	V6
Crew	3	4	4	4	4
Powerplant	2 x DB 606	2 x DB 606	2 x DB 606	2 x DB 606	2 x DB 606
Output (each)	2,600 hp	2,600 hp	2,600 hp	2,600 hp	2,700 hp
Length	20.58 m	20.58 m	20.58 m	20.58 m	20.58 m
	(67 ft 6 in)	(67 ft 6 in)	(67 ft 6 in)	(67 ft 6 in)	(67 ft 6 in)
Height	6.67 m	6.67 m	6.67 m	6.67 m	6.67 m
	(21 ft 10fi in)	(21 ft 10fi in)	(21 ft 10fi in)	(21 ft 10fi in)	(21 ft 10fi in)
Wing Span	31.40 m	31.40 m	31.40 m	31.40 m	31.40 m
	(103 ft 0 in)	(103 ft 0 in)	(103 ft 0 in)	(103 ft 0 in)	(103 ft 0 in)
Wing Area	100.00 m≈	100.00 m≈	100.00 m≈	100.00 m≈	100.00 m≈
	(1,076.40 ft≈)	(1,076.40 ft≈)	(1,076.40 ft≈)	(1,076.40 ft≈)	(1,076.40 ft≈)
Wing Loading	239.20 kg/m≈	241.20 kg/m≈	248.20 kg/m≈	249.50 kg/m≈	280.00 kg/m≈
	(49.00 lb/ft≈)	(49.40 lb/ft≈)	(50.83 lb/ft≈)	(51.10 lb/ft≈)	(57.34 lb/ft≈)
Power Loading	4.60 kg/hp	4.60 kg/hp	4.77 kg/hp	4.77 kg/hp	5.19 kg/hp
	(10.14 lb/hp)	(10.14 lb/hp)	(10.51 lb/hp)	(10.51 lb/hp)	(11.44 lb/hp)
Weight (empty)	13,720 kg	13,720 kg	14,220 kg	14,220 kg	16,730 kg
	(30,247 lb)	(30,247 lb)	(31,350 lb)	(31,350 lb)	(36,883 lb)
Weight (loaded)	23,920 kg	24,130 kg	24,820 kg	24,920 kg	28,000 kg
	(52,735 lb)	(53,197 lb)	(54,719 lb)	(54,939 lb)	(61,729 lb)
Useful Load	10,200 kg	10,410 kg	10,600 kg	10,700 kg	11,270 kg
	(22,487 lb)	(22,950 lb)	(23,369 lb)	(23,589 lb)	(24,846 lb)
Maximum Speed	460 km/h	460 km/h	460 km/h	460 km/h	465 km/h
	(286 mph)	(286 mph)	(286 mph)	(286 mph)	(290 mph)
Cruising Speed	410 km/h	410 km/h	410 km/h	425 km/h	425 km/h
	(255 mph)	(255 mph)	(255 mph)	(264 mph)	(264 mph)
Ceiling	7,000 m	7,000 m	7,000 m	7,000 m	7,000 m
	(22,966 ft)	(22,966 ft)	(22,966 ft)	(22,966 ft)	(22,966 ft)
Range	5,000 km	5,000 km	5,400 km	5,400 km	5,400 km
	(3,107 miles)	(3,107 miles)	(3,355 miles)	(3,355 miles)	(3,355 miles)
Armament					
Nose (upper)	1 x MG 131	1 x MG 15	1 x MG 131	1 x MG 131	1 x MG 131
Nose (lower)	1 x MG 131	1 x MG 15	1 x MG 131	1 x MG 131	1 x MG FF
Dorsal (fwd)	1 x MG 131	1 x MG 15	1 x MG 131	1 x MG 131	1 x MG FF
Dorsal (aft)	---	---	---	---	---
Ventral	---	---	---	---	---
Tail	1 x MG 131	1 x MG 15	1 x MG 131	1 x MG 131	1 x MG 131
Bomb-Load (min)	---	---	---	---	800 kg
	---	---	---	---	(1,764 lb)
Bomb-Load (max)	---	---	---	---	2,450 kg
	---	---	---	---	(5,401 lb)

V7	V8	V9	V14	V15	V19
4	5	5	5	5	5
2 x DB 606	2 x DB 606	2 x DB 606	2 x DB 610	2 x DB 610	2 x DB 610
2,700 hp	2,700 hp	2,700 hp	2,950 hp	2,950 hp	2,950 hp
20.40 m	20.40 m	20.40 m	20.40 m	20.40 m	20.40 m
(66 ft 11 in)	(66 ft 11 in)	(66 ft 11 in)	(66 ft 11 in)	(66 ft 11 in)	(66 ft 11 in)
6.67 m	6.67 m	6.67 m	6.67 m	6.67 m	6.67 m
(21 ft 10fi in)	(21 ft 10fi in)	(21 ft 10fi in)	(21 ft 10fi in)	(21 ft 10fi in)	(21 ft 10fi in)
31.40 m	31.40 m	31.44 m	31.40 m	31.44 m	31.44 m
(103 ft 0 in)	(103 ft 0 in)	(103 ft 1fl in)	(103 ft 0 in)	(103 ft 1fl in)	(103 ft 1fl in)
100.00 m≈	100.00 m≈	100.00 m≈	100.00 m≈	100.00 m≈	100.00 m≈
(1,076.40 ft≈)	(1,076.40 ft≈)	(1,076.40 ft≈)	(1,076.40 ft≈)	(1,076.40 ft≈)	(1,076.40 ft≈)
267.00 kg/m≈	279.10 kg/m≈	285.00 kg/m≈	300.00 kg/m≈	310.00 kg/m≈	270.00 kg/m≈
(54.70 lb/ft≈)	(57.12 lb/ft≈)	(58.40 lb/ft≈)	(61.44 lb/ft≈)	(63.50 lb/ft≈)	(55.30 lb/ft≈)
4.95 kg/hp	5.17 kg/hp	5.28 kg/hp	5.09 kg/hp	5.25 kg/hp	4.50 kg/hp
(10.91 lb/hp)	(11.40 lb/hp)	(11.64 lb/hp)	(11.22 lb/hp)	(11.57 lb/hp)	(9.92 lb/hp)
16,600 kg	17,210 kg	17,700 kg	18,040 kg	18,740 kg	16,800 kg
(36,597 lb)	(37,941 lb)	(39,021 lb)	(39,771 lb)	(41,315 lb)	(37,038 lb)
26,700 kg	27,910 kg	28,500 kg	30,000 kg	31,000 kg	27,000 kg
(58,863 lb)	(61,351 lb)	(62,832 lb)	(66,139 lb)	(68,343 lb)	(59,524 lb)
10,100 kg	10,700 kg	10,800 kg	11,960 kg	12,200 kg	10,200 kg
(22,267 lb)	(23,589 lb)	(23,810 lb)	(26,367 lb)	(26,896 lb)	(22,487 lb)
465 km/h	465 km/h	510 km/h	510 km/h	520 km/h	520 km/h
(290 mph)	(290 mph)	(317 mph)	(317 mph)	(323 mph)	(323 mph)
425 km/h	425 km/h	430 km/h	450 km/h	450 km/h	440 km/h
(264 mph)	(264 mph)	(267 mph)	(280 mph)	(280 mph)	(273 mph)
7,000 m	7,000 m	7,000 m	7,000 m	7,000 m	7,000 m
(22,966 ft)	(22,966 ft)	(22,966 ft)	(22,966 ft)	(22,966 ft)	(22,966 ft)
5,400 km	5,600 km	5,600 km	5,600 km	5,600 km	6,200 km
(3,355 miles)	(3,480 miles)	(3,480 miles)	(3,480 miles)	(3,480 miles)	(3,852 miles)
1 x MG 15	1 x MG 131	1 x MG 131	1 x MG 131	1 x MG 131	1 x MG 81
1 x MG 15	1 x MG 131	1 x MG 131	1 x MG 131	1 x MG FF	1 x MG 151
1 x MG 15	1 x MG FF	1 x MG 131	1 x MG 131	1 x MG 131	2 x MG 131
—-	—-	E2	E2	E2	E2
—-	—-	1 x MG 81Z	1 x MG 81Z	1 x MG 81Z	1 x MG 131
1 x MG 131	1 x MG 131	1 x MG 131	1 x MG 131	1 x MG 131	1 x MG 151
—-	800 kg	—-	800 kg	800 kg	800 kg
—-	(1,764 lb)	—-	(1,764 lb)	(1,764 lb)	(1,764 lb)
—-	2,450 kg	—-	2,450 kg	2,600 kg	2,800 kg
—-	(5,401 lb)	—-	(5,401 lb)	(5,732 lb)	(6,173 lb)

Appendix 2

He 177A-0 to A-7

	A-0	A-1	A-3/R1	A-3/R2	A-3/R5
Crew	5	5	5	5	5
Powerplant	2 x DB 606	2 x DB 606	2 x DB 606	2 x DB 606	2 x DB 610
Output (each)	2,700 hp	2,700 hp	2,700 hp	2,700 hp	2,950 hp
Length	22.00 m	22.00 m	22.00 m	22.00 m	22.00 m
	(72 ft 2∕in)	(72 ft 2∕in)	(72 ft 2∕in)	(72 ft 2∕in)	(72 ft 2∕in)
Height	6.67 m	6.67 m	6.67 m	6.67 m	6.67 m
	(21 ft 10fi in)	(21 ft 10fi in)	(21 ft 10fi in)	(21 ft 10fi in)	(21 ft 10fi in)
Wing Span	31.44 m	31.44 m	31.44 m	31.44 m	31.44 m
	(103 ft 1fl in)	(103 ft 1fl in)	(103 ft 1fl in)	(103 ft 1fl in)	(103 ft 1fl in)
Wing Area	100.00 m≈	100.00 m≈	100.00 m≈	100.00 m≈	100.00 m≈
	(1,076.40 ft≈)	(1,076.40 ft≈)	(1,076.40 ft≈)	(1,076.40 ft≈)	(1,076.40 ft≈)
Wing Loading	300.00 kg/m≈	303.00 kg/m≈	298.00 kg/m≈	298.00 kg/m≈	288.00 kg/m≈
	(61.44 lb/ft≈)	(62.05 lb/ft≈)	(61.03 lb/ft≈)	(61.03 lb/ft≈)	(59.00 lb/ft≈)
Power Loading	5.56 kg/hp	5.62 kg/hp	5.52 kg/hp	5.52 kg/hp	4.88 kg/hp
	(12.26 lb/hp)	(12.40 lb/hp)	(12.17 lb/hp)	(12.17 lb/hp)	(10.76 lb/hp)
Weight (empty)	17,000 kg	16,500 kg	16,600 kg	16,800 kg	18,410 kg
	(37,478 lb)	(36,376 lb)	(36,597 lb)	(37,037 lb)	(40,587 lb)
Weight (loaded)	30,000 kg	30,300 kg	29,800 kg	29,800 kg	29,890 kg
	(66,139 lb)	(66,800 lb)	(65,698 lb)	(65,698 lb)	(65,896 lb)
Useful Load	13,000 kg	13,800 kg	13,200 kg	13,000 kg	10,480 kg
	(28,660 lb)	(30,424 lb)	(29,101 lb)	(28,660 lb)	(23,104 lb)
Maximum Speed	480 km/h	480 km/h	480 km/h	480 km/h	480 km/h
	(298 mph)	(298 mph)	(298 mph)	(298 mph)	(298 mph)
Cruising Speed	410 km/h	410 km/h	410 km/h	410 km/h	410 km/h
	(255 mph)	(255 mph)	(255 mph)	(255 mph)	(255 mph)
Ceiling	10,000 m	10,000 m	10,000 m	10,000 m	10,000 m
	(32,808 ft)	(32,808 ft)	(32,808 ft)	(32,808 ft)	(32,808 ft)
Range	5,600 km	5,600 km	5,600 km	5,600 km	5,800 km
	(3,480 miles)	(3,480 miles)	(3,480 miles)	(3,480 miles)	(3,604 miles)
Armament					
Nose (upper)	1 x MG 81I	1 x MG 81I	1 x MG 81I	1 x MG 81I	1 x MG 81I
Nose (lower)	1 x MG FF	1 x MG FF	1 x MG FF	2 x MG 151	—
Dorsal (fwd)	2 x MG 131	2 x MG 131	2 x MG 131	2 x MG 131	2 x MG 131
Dorsal (aft)	—	—	1 x MG 131	1 x MG 131	—
Ventral	1 x MG 81Z	1 x MG 131	1 x MG 131	1 x MG 131	—
Tail	1 x MG 131	1 x MG 131	1 x MG 131	1 x MG 151	1 x MG 151
Bomb-Load (min)	800 kg	800 kg	800 kg	800 kg	1,600 kg
	(1,764 lb)	(1,764 lb)	(1,764 lb)	(1,764 lb)	(3,527 lb)
Bomb-Load (max)	2,200 kg	2,200 kg	2,200 kg	2,500 kg	2,500 kg
	(4,850 lb)	(4,850 lb)	(4,850 lb)	(5,512 lb)	(5,512 lb)

A-5/R2	A-5/R7	A-6/R1	A-6/R2	A-7
6	6	6	6	6
2 x DB 610	2 x DB 610	2 x DB 610	2 x DB 610	2 x DB 610
2,975 hp	2,975 hp	2,975 hp	2,975 hp	2,975 hp
22.00 m	22.00 m	22.00 m	22.00 m	22.00 m
(72 ft 2/in)	(72 ft 2/in)	(72 ft 2/in)	(72 ft 2/in)	(72 ft 2/in)
6.67 m	6.67 m	6.67 m	6.67 m	6.67 m
(21 ft 10fi in)	(21 ft 10fi in)	(21 ft 10fi in)	(21 ft 10fi in)	(21 ft 10fi in)
31.44 m	31.44 m	31.44 m	31.44 m	31.44 m
(103 ft 1fl in)	(103 ft 1fl in)	(103 ft 1fl in)	(103 ft 1fl in)	(103 ft 1fl in)
100.00 m≈	100.00 m≈	100.00 m≈	100.00 m≈	108.00 m≈
(1,076.40 ft≈)	(1,076.40 ft≈)	(1,076.40 ft≈)	(1,076.40 ft≈)	(1,162.50 ft≈)
310.00 kg/m≈	310.00 kg/m≈	320.00 kg/m≈	326.00 kg/m≈	320.00 kg/m≈
(63.50 lb/ft≈)	(63.50 lb/ft≈)	(65.54 lb/ft≈)	(66.76 lb/ft≈)	(65.54 lb/ft≈)
5.25 kg/hp	5.21 kg/hp	5.38 kg/hp	5.50 kg/hp	5.82 kg/hp
(11.57 lb/hp)	(11.49 lb/hp)	(11.86 lb/hp)	(12.12 lb/hp)	(12.83 lb/hp)
16,800 kg	17,000 kg	16,800 kg	16,900 kg	18,100 kg
(37,038 lb)	(37,478 lb)	(37,038 lb)	(37,258 lb)	(39,904 lb)
32,000 kg	32,000 kg	32,000 kg	32,600 kg	34,600 kg
(70,548 lb)	(70,548 lb)	(70,548 lb)	(71,871 lb)	(76,280 lb)
15,200 kg	15,000 kg	15,200 kg	15,700 kg	16,500 kg
(33,510 lb)	(33,069 lb)	(33,510 lb)	(34,612 lb)	(36,376 lb)
440 km/h	520 km/h	520 km/h	535 km/h	540 km/h
(273 mph)	(323 mph)	(323 mph)	(332 mph)	(336 mph)
380 km/h	440 km/h	440 km/h	440 km/h	440 km/h
(236 mph)	(273 mph)	(273 mph)	(273 mph)	(273 mph)
8,000 m	15,200 m	15,000 m	15,000 m	16,000 m
(26,246 ft)	(49,869 ft)	(49,213 ft)	(49,213 ft)	(52,493 ft)
6,000 km	6,000 km	5,800 km	5,800 km	7,200 km
(3,728 miles)	(3,728 miles)	(3,604 miles)	(3,604 miles)	(4,474 miles)
1 x MG 81I	1 x MG 81I	1 x MG 81I	—	1 x MG 81I
1 x MG 151	1 x MG 151	1 x MG 151	1 x MG 131Z	1 x MG 151
2 x MG 131	2 x MG 131	2 x MG 131	1 x FDL 151Z	1 x MG 131
1 x MG 131	1 x MG 131	—	—	—
1 x MG 131	—	—	1 x MG 131	—
1 x MG 151	1 x MG 151	1 x HL 131V	1 x HL 131V	1 x MG 151
2,800 kg	2,800 kg	2,500 kg	2,500 kg	2,500 kg
(6,173 lb)	(6,173 lb)	(5,512 lb)	(5,512 lb)	(5,512 lb)
2,800 kg	2,800 kg	3,500 kg	3,500 kg	4,200 kg
(6,173 lb)	(6,173 lb)	(7,716 lb)	(7,716 lb)	(9,259 lb)

Appendix 3

Daimler-Benz DB 606/610/613 Aircraft Powerplants

Model	Manufactured from	Coupled powerplant consisting of
DB 606V	February 1937	2 x DB 601V

2 x 12-cylinder injection aircraft engines with supercharger, common reduction gearing and separate cut-off devices for each engine.

DB 606A-0/-1	February 1937	2 x DB 601A/E
DB 606B-1	August 1938	2 x DB 601A/E

B4 aviation fuel.

DB 606C-1	August 1938	2 x DB 601A
DB 606D-1	August 1938	2 x DB 601A

C3 aviation fuel. Power output: 2,600-2,700 hp at 2,700 rpm at sea level; cubic capacity: 68 ltr (15 Imp gal).

DB 610V	June 1940	2 x DB 605V

2 x 12-cylinder injection aircraft engines (similar type of construction to DB 601 in DB 606 coupled unit).

DB 610A-1	November 1940	2 x DB 605A
DB 610B-1	November 1940	2 x DB 605B

B4 aviation fuel.

DB 610B-2	1944	2 x DB 605D

Coupled powerplant with DB 606 reduction gear; B4 aviation fuel.

DB 610BS	1944	2 x DB 605WS/XS

Coupled powerplant with DB 603G supercharger; C2/C3 aviation fuel.

DB 610C-1	February 1942	2 x DB 605D

Coupled powerplant with higher compression ratio; C2/C3 aviation fuel.

DB 610C-2	1943	2 x DB 605D

Coupled powerplant with improved DB 605D engines; C2/C3 aviation fuel.

DB 610D-1	February 1942	2 x DB 605D
DB 610D-2	1943	2 x DB 605D

Progressive development of DB 610B-1/-2; C2/C3 aviation fuel.

DB 613	March 1940	2 x DB 603V

2 x 12-cylinder injection aircraft engines (similar type of construction to DB 601 engines in DB 606 coupled unit).

DB 613A-0	1941	2 x DB 603W/X

Pre-production series.

DB 613A-1	1941	2 x DB 603E

Coupled powerplant with 7,100 m (23,294 ft) rated altitude; B4 aviation fuel.

DB 613B-0	1943	2 x DB 603B

Coupled powerplant with higher compression ratio; C3 aviation fuel.

DB 613B-1	1944	2 x DB 603E

Refined version of DB 613B-0; C3 aviation fuel.

DB 613C-0	July 1942	2 x DB 603A

Coupled powerplant with higher compression ratio; C3 aviation fuel.

DB 613D-0	July 1942	2 x DB 603A/F

Coupled powerplant with DB 606A reduction gear.

DB 613D-1	1941	2 x DB 603E-1

Coupled powerplant with DB 610 reduction gear.

Appendix 4

Heinkel He 177 Prototypes

V1 (CB+RP/Wk-Nr 00 0001):
Flight characteristics and performance tests; evaluation of Fowler flaps. Severely damaged as result of crash-landing (3 October 1941).

V2 (CB+RQ/Wk-Nr 00 0002):
Stability tests and evaluation of new horizontal tailplane. Destroyed in crash 27 June 1940.

V3 (D-AGIG/Wk-Nr 00 0003):
Stability tests and evaluation of new horizontal tailplane. Destroyed in crash 24 April 1940 due to tailplane failure.

V4 (Wk-Nr 00 0004):
Performance and inclined flight tests; evaluation of brake flaps. Destroyed in crash 8 June 1941.

V5 (PM+OD/Wk-Nr 00 0005):
Performance and inclined flight tests; tailplane experiments. To Heinkel works 23 June 1943.

V6 (BC+BP/Wk-Nr 00 0006):
First armament pattern aircraft. Assigned to KG 40 as training aircraft 23 June 1943.

V7 (SF+TB/Wk-Nr 00 0007):
Second armament pattern aircraft. Assigned to KG 40 as training aircraft 23 June 1943.

V8 (SF+TC/Wk-Nr 00 0008):
Evaluation of flaps, radiators, shortened Fowler flaps, powerplants. At Heinkel works 26 June 1943.

V9 (GA+QQ/Wk-Nr 00 0023):
A-08 rebuilt to B-5 standard. Evaluation of four separate engines, twin fin/rudder assembly and K-12 automatic flight control system. Handed over to *E-Stelle* Rechlin 15 August 1944.

V10 (GA+QR/Wk-Nr 00 0024):
A-09 rebuilt to A-2 standard. Evaluation of pressurised cockpit and K-12 automatic flight control system. Handed over to Heinkel training workshops 28 June 1944; cannibalised.

V11 (GA+QS/Wk-Nr 00 0025):
A-010 rebuilt to A-2 standard. Evaluation of pressurised cockpit, strengthened armament and K4Ü device. Handed over to *E-Stelle* Rechlin 26 June 1944.

V12 (GI+BL/Wk-Nr 15151):
A-1 pattern aircraft for current/consecutive alterations and armament. Handed over to Eger depot 22 July 1943.

V13 (identity unknown):
Planned as replacement for A-0 series.

V14 (GI+BM/Wk-Nr 15152):
A-1 experimental aircraft for *Kehl III* equipment. Final whereabouts unknown.

V15 (Wk-Nr 355001):
A-3 fitted with new cockpit and used for inclined flight trials with brake parachute. Crashed 24 June 1944.

V16 (Wk-Nr 355004):
A-3 pattern aircraft for radio equipment; delivered 28 October 1942.

V17 (Wk-Nr 355005):
A-3 pattern aircraft with aerial torpedo installation; delivered 14 April 1942.

V18 (GA+QX/Wk-Nr 00 0030):
A-015; first experimental *Zerstörer* aircraft with 30 mm MK 101 cannon installation in lower nose position.

V19 (VF+QA/Wk-Nr 332101):
A-3 used for DB 610 powerplant tests. Destroyed as result of engine fire 13 November 1943.

V20 (VF+RD/Wk-Nr 15254):
A-1 rebuilt to A-3 standard. Fitted with flexible gun installations and MG 131Z tail turret. Aircraft still not serviceable as of 11 February 1944.

V21 (VF+QB/Wk-Nr 332102):
A-3 rebuilt to A-5 standard. Pattern aircraft for A-5 series with *Kehl III/IV* equipment. At Heinkel as of 2 March 1944.

V22 (VF+QD/Wk-Nr 332104):
A-3 rebuilt as first A-5 pattern aircraft. Used to evaluate stall characteristics. Assigned to de-icing trials at Munich-Riem December 1943.

V23 (VF+QL/Wk-Nr 332112):
A-3 rebuilt to A-5 standard. Used to evaluate performance, slotted ailerons, Fowler flaps and DB 610s. Handed over to FFS (B) 16 at Brandis June 1944.

V24 (ND+SS/Wk-Nr 135024):
A-3 rebuilt to A-5 standard. Used for de-icing equipment and DB 610 trials. Handed over to *E-Stelle* Rechlin 6 November 1943.

V25 (GI+BN/Wk-Nr 15153):
A-1 used as pattern aircraft for enlarged fin/rudder and modified undercarriage. Handed over to *E-Stelle* Rechlin 28 April 1944.

V26 (GA+QM/Wk-Nr 00 0019):
A-04; used as replacement aircraft with improved aerial torpedo installation.

V27 (Wk-Nr 15203):
A-1 with enlarged tailplane and shortened fuselage. In use as test aircraft October 1943. To 1./KG 1 as V4+UC. Destroyed in crash 5 April 1944.

V28 (Wk-Nr 355056):
A-3 used for trials of automatic flight controls. Handed over to Patin company 22 July 1943.

V29 (GO+IF/Wk-Nr 15155):
A-1 pattern aircraft for remote-controlled nose gun installation. Still in experimental service as of October 1943. Burned out.

V30 (ND+SM/Wk-Nr 135018):
A-3 pattern aircraft for improved aerial torpedo installation. On trials work 1943-44.

V31 (TM+IF/Wk-Nr 550202):
A-5 trials aircraft for LT 950 (L 10/LT 10) aerial torpedo equipment. At *E-Stelle* Gotenhafen-Hexengrund in 1944 for trials with aerial torpedoes.

V32 (GP+WC/Wk-Nr 535353):
A-3 pattern aircraft for quadruple-barrel HL 131V tail turret July 1943.

V33 (GP+WD/Wk-Nr 535354):
A-3 pattern aircraft for quadruple-barrel HL 131V tail turret July 1943.

V34 (GP+WN/Wk-Nr 535364):
A-3 pattern aircraft for quadruple-barrel HL 131V tail turret July 1943.

V35 (GP+WP/Wk-Nr 535366):
A-3 used for evaluation of pressurised cockpit and inclined flight tests. Handed over to *E-Stelle* Werneuchen July 1943.

V36 (Wk-Nr 332121):
A-1 brought up to A-3 standard. Used for wing de-icing trials. Fitted with improved *Kehl IV* equipment. Crash-landed 14 January 1944.

V37 (KM+TN/Wk-Nr 550038):
A-5 pattern aircraft for FuG 207 *Neptun* and 208 radars. Fitted with underfuselage ETC bomb rack. Handed over to *E-Stelle* Peenemünde 25 May 1944.

V38 (KM+TB/Wk-Nr 550002):
A-5 pattern aircraft fitted with FuG 200 *Hohentwiel* and 208 radars; two MG 131s in nose gun position. Handed over to *E-Stelle* Werneuchen 27 April 1944. Remains captured at Prague-Rusiye 8 May 1945.

V39 (KM+TY/Wk-Nr 550049):
A-5 used for inclined flight tests and to evaluate PDS installation. Fitted with wooden bomb-bay doors. Handed over to *E-Stelle* Rechlin July 1944.

V40 (GR+MH/Wk-Nr 535850):
A-3 pattern aircraft for FuG 217 and *Berlin* radars. Handed over to *E-Stelle* Werneuchen 25 July 1944.

V101 (NN+QQ/Wk-Nr 535550):
A-3 rebuilt to B-5 standard. Four separate engines, lateral paddles; used for performance determination. On hand as trials aircraft August 1944; scrapped.

V102 (GA+QQ/Wk-Nr 00 0023):
A-08/V9 (qv) rebuilt to B-5 standard and redesignated V102. Four separate engines, twin fin/rudder assembly. Used for stability tests. Handed over to *E-Stelle* Rechlin August 1944; scrapped.

V103 (KM+TL/Wk-Nr 550036):
A-5 rebuilt to B-5 standard. Four separate engines, quadruple-barrel tail turret. On hand as test aircraft (June 1944); destroyed in air raid July 1944.

V104 (KM+TE/Wk-Nr 550005):
A-5 rebuilt to B-5 standard. Pattern aircraft for He 177B-5 series. Believed destroyed in air raid before completion July 1944.

Appendix 5

He 177 Production Series

Model	Role	Powerplants	Defensive Armament	
A-0	Heavy bomber	DB 606A/B	Nose (upper)	1 x MG 81I
			Nose (lower)	1 x MG FF
			Dorsal (fwd)	1 X MG 81
			or	
			1 x MG 131	
			Ventral	1 x MG 81Z
			Tail	1 x MG 131A1

Pre-production series aircraft with different defensive armament.

A-1/R1	Heavy bomber	DB 606A/B	Nose (upper)	1 x MG 81I
			Nose (lower)	1 x MG FF
			Dorsal (fwd)	1 x FDL B 131/1A
			Ventral	1 x MG 81Z
			Tail	1 x MG 131A1
A-1/R2	Heavy bomber	DB 606A/B	Nose (upper)	1 x MG 81I
			Nose (lower)	1 x MG FF
			Dorsal (fwd)	1 x FDL B 131/1A
			Ventral (aft)	1 x FDL C 131/1A
			Tail	1 x MG 131A1

Experimental model with aft ventral gun position located behind the bomb-bay.

A-1/R3	Heavy bomber	DB 606A/B	Nose (upper)	1 x MG 81I
			Nose (lower)	1 x MG FF
			Dorsal (fwd)	1 x FDL B 131/1A
			Ventral (aft)	1 x FDL C 131/1A
			Tail	1 x MG 131A1

As per A-1/R2 except for reduced ammunition supply to upper nose gun position.

A-1/R4	Heavy bomber	DB 606A/B	Nose (upper)	1 x MG 81I
			Nose (lower)	1 x MG FF
			Dorsal (fwd)	1 x FDL B 131/1A
			Dorsal (aft)	1 x MG 131
			Ventral	1 x MG 131
			Tail	1 x MG 131A1

Final He 177A-1 sub-model.

A-1/U2	Long-range *Zerstörer*	DB 606A/B	Nose (upper)	1 x MG 81I
			Nose (lower)	2 x MK 101
			Dorsal (fwd)	1 x FDL B 131/1A
			Tail	1 x MG 131A1

Series rebuild of 12 A-1 bombers.

A-2	High-altitude bomber	DB 606	Nose (upper)	1 x FL 81Z
			Dorsal (fwd)	1 x FL 81Z
			Ventral	1 x FL 81Z
			Tail	1 x MG 131A1

Three-seat pressurised cabin and pressurised tail gun position.

A-3/R1	Heavy bomber	DB 606A/B	Nose (upper)	1 x MG 81I
		or	Nose (lower)	1 x MG FF
		DB 610A/B	Dorsal (fwd)	1 x FDL B 131/2A
			Dorsal (aft)	1 x MG 131
			Ventral	1 x MG 131
			Tail	1 x MG 131A1

Model	Role	Powerplants	Defensive Armament	

New airframe with A-1/R4 armament configuration, but with remote-controlled twin-barrel turrets.

A-3/R2	Heavy bomber	DB 610A/B	Nose (upper)	1 x MG 81I
			Nose (lower)	1 x MG FF
			Dorsal (fwd)	1 x FDL B 131/2A
			Dorsal (aft)	1 x MG 131
			Ventral	1 x MG 131
			Tail	1 x MG 151/20

Sub-model with more powerful tail gun.

A-3/R3	Glide-bomb carrier	DB 610A/B	Nose (upper)	1 x MG 81I
			Nose (lower)	1 x MG FF
			Dorsal (fwd)	1 x FDL B 131/2A
			Dorsal (aft)	1 x MG 131
			Ventral	1 x MG 131
			Tail	1 x MG 151/20

Three attachment points (two underwing, one underfuselage) for Hs 293 glide-bombs. Fitted with FuG 203 radio guidance equipment.

A-3/R4	Glide-bomb carrier	DB 610A/B	Nose (upper)	1 x MG 81I
			Nose (lower)	1 x MG FF
			Dorsal (fwd)	1 x FDL B 131/2A
			Dorsal (aft)	1 x MG 131
			Ventral	1 x MG 131
			Tail	1 x MG 151/20

Three attachment points (two underwing, one underfuselage) for Hs 293 glide-bombs. Fitted with *Kehl III* radio guidance equipment.

A-3/R5	*Zerstörer*	DB 610A/B	Nose (upper)	1 x MG 81I
			Nose (lower) (PaK 40L)	1 x BK 7,5
			Dorsal (fwd)	1 x FDL B 131/2A
			Dorsal (aft)	1 x MG 131
			Tail	1 x MG 151/20

Project study only.

A-3/R6	Heavy bomber	DB 610A/B	Nose (upper)	1 x MG 81I
			Nose (lower)	1 x MG FF
			Dorsal (fwd)	1 x FDL B 131/2A
			Dorsal (aft)	1 x MG 131
			Ventral	1 x MG 131
			Tail	1 x MG 151/20

A-3 optimised for long-range operations; bomb-load in rear bomb-bay only.

A-3/R7	Torpedo carrier	DB 610A/B	Nose (upper)	1 x MG 81I
			Nose (lower)	1 x MG FF
			Dorsal (fwd)	1 x FDL B 131/2A
			Dorsal (aft)	1 x MG 131
			Ventral	1 x MG 131
			Tail	1 x MG 131A1

Four ETCs (two underwing, two underfuselage) for LT 5 aerial torpedoes. Defensive armament as per A-3/R1 or reduced.

A-4	High-altitude bomber	DB 610A/B	Nose (upper)	1 x FL 81Z
			Dorsal (fwd)	1 x FL 81Z
			Ventral	1 x FL 81Z
			Tail	1 x MG 131A1

Model	Role	Powerplants	Defensive Armament	

Project study only. Defensive armament as per A-2 or increased.

A-5/R1	Heavy bomber	DB 610A/B	Nose (upper)	1 x MG 81I
			Nose (lower)	1 x MG 151/20
			Dorsal (fwd)	1 x FDL B 131/2A
			Dorsal (aft)	1 x MG 131
			Ventral	1 x MG 81Z
			Tail	1 x MG 151/20

Carrier aircraft for Hs 293A-1/B-1 and PC 1400X glide-bombs.

A-5/R2	Heavy bomber	DB 610A/B	Nose (upper)	1 x MG 81I
			Nose (lower)	1 x MG 151/20
			Dorsal (fwd)	1 x FDL B 131/2A
			Dorsal (aft)	1 x MG 131
			Ventral	1 x MG 131
			Tail	1 x MG 151/20

Strengthened ventral armament; modified bomb-release equipment for use in conjunction with three-section bomb-bay.

A-5/R3	Heavy bomber	DB 610A/B	Nose (upper)	1 x MG 81I
			Nose (lower)	1 x MG 151/20
			Dorsal (fwd)	1 x FDL B 131/2A
			Dorsal (aft)	1 x MG 131
			Ventral (fwd)	1 x MG 131
			Ventral (aft)	1 x FDL C 1312/1A
			Tail	1 x MG 151/20

Strengthened defensive armament.

A-5/R4	Heavy bomber	DB 610A/B	Nose (upper)	1 x MG 81I
			Nose (lower)	1 x MG 151/20
			Dorsal (fwd)	1 x FDL B 131/2A
			Dorsal (aft)	1 x MG 131
			Ventral	1 x MG 81Z
			Tail	1 x MG 151/20

Project incorporating *Kehl* radio-guidance equipment and simplified bomb-release mechanism.

A-5/R5	Heavy bomber	DB 610A/B	Nose (upper)	1 x MG 81I
			Nose (lower)	1 x MG 151/20
			Dorsal (fwd)	1 x FDL B 131/2A
			Dorsal (aft)	1 x MG 131
			Ventral (fwd)	1 x MG 81Z
			Ventral (aft)	1 x FDL C 131/1A
			Tail	1 x MG 151/20

Strengthened defensive armament.

A-5/R6	Heavy bomber	DB 610A/B	Nose (upper)	1 x MG 81I
			Nose (lower)	1 x MG 151/20
			Dorsal (fwd)	1 x FDL B 131/2A
			Dorsal (aft)	1 x MG 131
			Ventral	1 x MG 81Z
			Tail	1 x MG 151/20

Smaller bomb-bay.

A-5/R7	Heavy bomber	DB 610A/B	Nose (upper)	1 x FL 81Z
			Dorsal (fwd)	1 x FL 81Z
			Ventral	1 x FL 81Z
			Tail	1 x MG 131A1

Model	Role	Powerplants	Defensive Armament	

Project study only. Pressurised cockpit.

A-6/R1	Heavy bomber	DB 610A/B	Nose (lower)	1 x FDL 131Z
			Dorsal (fwd)	1 x FDL 151Z
			Ventral	1 x WL 131Z
			Tail	1 x HL 131V

Pressurised cockpit defensive armament; smaller bomb-bay.

A-6/R2	Heavy bomber	DB 610A/B	Nose (lower)	1 x DL 131Z
			Dorsal (fwd)	1 x DL 151Z
			Ventral	1 x MG 131
			Tail	1 x HDL 81V

New cockpit form with remote-controlled lower nose gun. Fitted with A-6/R1's smaller bomb-bay.

A-7	Heavy bomber	DB 610A/B	Nose (lower)	1 x FDL 131Z
		or	Dorsal (fwd)	1 x FDL 151Z
		DB 613A/B	Ventral	2 x WL 131Z
			Tail	1 x HL 131V

Increased wing span and area.

A-8	Heavy bomber	BMW 801E		
		Jumo 213		
		or		
		DB 603		

Rebuild of A-5 with four separate engines.

A-10	Heavy bomber	BMW 801E	Nose (lower)	1 x FDL 131Z
			Dorsal (fwd)	1 x FDL 151Z
			Ventral	2 x WL 131Z
			Tail	1 x HL 131V

Rebuild of A-7.

B-5/R1	Heavy bomber	DB 603A	Nose (lower)	1 x BL 131Z
			Dorsal (fwd)	1 x MG 151Z
			Dorsal (aft)	1 x MG 151Z
			Ventral	1 x MG 131I
			Tail	1 x HL 131V

Rebuild of A-5.

B-5/R2	Heavy bomber	DB 603A	Nose (lower)	1 x BL 131V
			Dorsal (fwd)	1 x MG 151Z
			Dorsal (aft)	1 x MG 151Z
			Ventral	1 x MG 131I
			Tail	1 x HL 131V

Project study only. Strengthened nose armament.

B-6/R1	Heavy bomber	DB 603A	Nose (lower)	1 x MG 151Z
			Dorsal (fwd)	1 x MG 151Z
			Dorsal (aft)	1 x MG 151Z
			Ventral	1 x MG 131I
			Tail	1 x HL 131V

Project study only. Development suspended August 1943.

B-6/R2	Heavy bomber	Jumo 213F	Nose (lower)	1 x MG 151Z
			Dorsal (fwd)	1 x MG 151Z
			Dorsal (aft)	1 x MG 151Z
			Ventral	1 x MG 131I
			Tail	1 x HL 131V

Model	Role	Powerplants	Defensive Armament	

Project study only. Defensive armament as per B-6/R1, but with additional Armament Set planned.

B-6/R3	Heavy bomber	Jumo 213F	Nose (lower)	1 x MG 151Z
			Dorsal (fwd)	1 x MG 151Z
			Dorsal (aft)	1 x MG 151Z
			Ventral	1 x MG 131I
			Tail	1 x HL 131V

Project study with greater fuel tankage and smaller bomb-bay.

B-7	Heavy bomber	Jumo 213F	Nose (lower)	1 x DL 131V
		Jumo 222A	Dorsal (fwd)	1 x FDL 151Z
		or	Ventral	1 x FDL 151Z
		DB 603E	Tail	1 x HL 131V

A-7 wings; improved cabin and strengthened undercarriage.

Appendix 6
Heinkel He 177

Werk-Nummer	Model	Codes	Last Mentioned	User/Use	Fate
000001	V1	CB+RP	03/10/41	1st prototype	Crashed
000002	V2	CB+RQ	27/06/40	2nd prototype	Crashed
000003	V3	D-AGIG	24/04/40	3rd prototype	Crashed
000004	V4	??+??	08/06/41	4th prototype	Ground contact
000005	V5	PM+OD	23/06/43	5th prototype	Wrecked
000006	V6	BC+BP	28/06/43	6th prototype	Belly-landed
000007	V7	SF+TB	28/06/43	7th prototype	Unknown
000008	V8	SF+TC	28/06/43	8th prototype	Unknown
000016	A-01	DL+AP	08/07/41	EHAG	TOA
000017	A-02	DL+AQ	28/06/43	KdE	Unknown
000018	A-03	DL+AR	12/11/41	EHAG, KdE	Unknown
000019	A-04	GA+QM	24/04/43	V26, KdE	Crashed
000020	A-05	GA+QN	13/06/42	KdE	Crashed (EF)
000021	A-06	GA+QO	20/02/42	KdE	Engine failure
000022	A-07	GA+QP	10/02/42	ESt 177	Burned out
000023	A-08	GA+QQ	28/06/43	V9, V102	Scrapped
000024	A-09	GA+QR	20/06/44	V10	Cannibalised
000025	A-010	GA+QS	26/06/44	V11	Unknown
000026	A-011	GA+QT	17/02/42	KdE	Crashed (EF)
000027	A-012	GA+QU	14/10/42	ESt 177	Damaged
000028	A-013	GA+QV	16/07/42	ESt 177	Crashed
000029	A-014	GA+QW	04/07/42	ESt 177	TOA
000030	A-015	GA+QX	28/06/43	V18 (*Zerstörer*)	Unknown
000031	A-016	GA+QY	Unknown	EHAG	Scrapped
000032	A-017	GA+QZ	06/05/42	ESt 177	Landing accident
000033	A-018	??+??	15/04/42	EHAG	Unknown
000034	A-019	??+??	16/04/42	KdE	Total loss
000035	A-020	??+??	16/04/42	ESt 177	Damaged
32001	A-0	DR+IJ	Unknown	EHAG	Damaged
32002	A-0	DR+IK	17/08/43	11./KG 40	Unknown
32003	A-0	DR+IL	06/11/42	IV/KG 40	Damaged
32004	A-0	F8+DV	03/02/44	11./KG 40	Unknown
32011	A-0	DR+IS	30/11/43	KdE	Unknown
32012	A-0	DR+IT	Unknown	IV/KG 40	Damaged
32013	A-0	DR+IU	??/09/44	IV/KG 40	War booty
15151	A-1	GI+BL	22/07/43	V12 (armament)	Unknown
15152	A-1	GI+BM	??/05/43	V14 (*Kehl III*)	Unknown
15153	A-1	GI+BN	28/04/44	V25, ESt 177	Unknown
15154	A-1	GI+BO	08/02/42	KdE	Engine fire
15155	A-1	GI+BP	??/05/43	V29 (gun trials)	Burned out
15156	A-1	GI+BQ	??/05/43	I/FKG 50	Unknown
15157	A-1	GI+BR	09/10/43	*Zerstörer*	Total loss
15158	A-1	GI+BS	21/02/42	Arado	Wing fracture
15159	A-1	GI+BT	??/05/43	2nd *Kehl III* a/c	Unknown
15160	A-1	GI+BU	??/05/43	*Zerstörer*	Unknown
15161	A-1	GI+BV	02/09/42	Long-range recce	Crashed
15162	A-1	GI+BW	28/05/43	3rd *Kehl III* a/c	Unknown
15163	A-1	GI+BX	??/05/43	*Zerstörer*	Unknown
15164	A-1	GI+BY	23/05/42	KG 100	Crashed
15165	A-1	GI+BZ	??/05/43	*Zerstörer*	Unknown

Werk-Nummer	Model	Codes	Last Mentioned	User/Use	Fate
15166	A-1	??+??	Unknown	I/KG 40	Unknown
15167	A-1	??+??	Unknown	I/KG 40	Unknown
15168	A-1	??+??	??/05/43	I/KG 40	Unknown
15169	A-1	??+??	??/05/43	KG 100	Unknown
15170	A-1	??+??	??/05/43	*Zerstörer*	Unknown
15171	A-1	BL+FA	15/05/43	KdE, ESt 177	Engine failure
15172	A-1	BL+FB	14/04/42	Arado	Engine failure
15173	A-1	BL+FC	??/06/43	*Kehl* trainer	Unknown
15174	A-1	BL+FD	Unknown	*Kehl* trainer	Unknown
15175	A-1	BL+FH	15/07/43	I/KG 40	Unknown
15176	A-1	BL+FF	Unknown	*Kehl* trainer	Unknown
15177	A-1	BL+FG	Unknown	TS 2	Unknown
15178	A-1	BL+FH	Unknown	I/KG 40	Unknown
15179	A-1	BL+FI	??/05/43	*Zerstörer*	Unknown
15180	A-1	BL+FJ	29/01/43	I/FKG 50	Crashed
15181	A-1	BL+FK	Unknown	*Kehl* trainer	Unknown
15182	A-1	BL+FL	Unknown	I/KG 40	Unknown
15183	A-1	BL+FM	??/05/43	*Kehl* trainer	Unknown
15184	A-1	BL+FN	??/05/43	*Kehl* trainer	Unknown
15185	A-1	BL+FO	??/ /43	*Kehl* trainer	Unknown
15186	A-1	BL+FP	??/05/43	*Kehl* trainer	Unknown
15187	A-1	BL+FQ	??/05/43	*Kehl* trainer	Unknown
15188	A-1	BL+FR	??/05/43	*Kehl* trainer	Unknown
15189	A-1	BL+FS	??/05/43	*Kehl* trainer	Unknown
15190	A-1	BL+FT	08/02/42	KdE, trials	Unknown
15191	A-1*	BL+FU	02/02/43	DVL, trials	Crashed
15192	A-1	BL+FV	14/05/43	DB, trials	Unknown
15193	A-1	BL+FW	??/05/43	*Kehl* trainer	Unknown
15194	A-1	BL+FX	??/05/43	*Kehl* trainer	Unknown
15195	A-1	BL+FY	25/01/43	KdE, trials	Unknown
15196	A-1	BL+FZ	??/05/43	*Kehl* trainer	Unknown
15197	A-1	DH+CW	15/03/44	2./KG 1	Crashed
15198	A-1	DH+CX	??/05/43	I/KG 1	Unknown
15199	A-1*	DH+CY	??/05/43	I/KG 1	Crashed
15200	A-1	DH+CZ	??/05/43	*Kehl* trainer	Unknown
15201	A-1	VE+UA	??/05/43	*Kehl* trainer	Unknown
15202	A-1	VE+UB	??/02/44	II/KG 40	Unknown
15203	A-1*	V4+UC	05/04/44	V27, 1./KG 1	Crashed
15204	A-1	VE+UD	??/05/43	*Kehl* trainer	Unknown
15205	A-1	VE+UE	??/05/43	*Kehl* trainer	Unknown
15206	A-1	VE+UF	??/05/43	*Kehl* trainer	Unknown
15207	A-1	VE+UG	23/11/43	IV/KG 40	Crashed
15208	A-1	VE+UH	03/04/43	4./KG 1	TOA
15209	A-1	VE+UI	??/05/43	*Kehl* trainer	Unknown
15210	A-1	VE+UJ	Unknown	Long-range recce	Unknown
15211	A-1	VE+UK	Unknown	Long-range recce	Unknown
15212	A-1	VE+UL	Unknown	Long-range recce	Unknown
15213	A-1	VE+UM	Unknown	ESt 177	Unknown
15214	A-1	VE+UN	28/05/43	KdE	Total loss
15215	A-1	VE+UO	21/12/42	ESt 177	Crashed
15216	A-1	VE+UP	Unknown	KdE, (diving)	Wrecked
15217	A-1	VE+UQ	??/05/43	KdE, Rechlin	Unknown
15218	A-1	VE+UR	14/01/43	KdE, DB trials	Unknown
15219	A-1	VE+US	Unknown	KdE, trials	Unknown

Werk-Nummer	Model	Codes	Last Mentioned	User/Use	Fate
15220	A-1	VE+UT	05/07/43	Unknown	Crash-landed
15221	A-1	VE+UU	Unknown	EHAG, I/KG 40	Unknown
15222	A-1	VE+UV	10/04/44	I/FKG 50, FFS (B) 15	Unknown
15223	A-1	VE+UW	Unknown	I/FKG 50	Unknown
15224	A-1	VE+UX	Unknown	I/FKG 50	Unknown
15225	A-1	VE+UY	Unknown	I/FKG 50	Landing accident
15226	A-1	VD+UA	24/06/44	FFS (B) 31	Unknown
15227	A-1	VD+UB	??/05/43	Trainer	Unknown
15228	A-1	VD+UC	??/06/43	KdE	Unknown
15229	A-1	VD+UD	??/05/43	Trainer	Pilot error
15230	A-1	VD+UE	Unknown	I/FKG 50	Crash-landed
15231	A-1	VD+UF	13/11/42	KdE	Total loss
15232	A-1	VD+UG	14/11/42	KdE	Total loss
15233	A-1*	E8+FH	16/01/43	I/FKG 50	MIA
15234	A-1	VD+UI	Unknown	I/FKG 50	Unknown
15235	A-1	VD+UJ	Unknown	I/FKG 50	Unknown
15236	A-1	VD+UK	30/01/43	I/FKG 50	Unknown
15237	A-1	VD+UL	Unknown	I/FKG 50	Unknown
15238	A-1	VD+UM	Unknown	I/FKG 50	Unknown
15239	A-1	VD+UN	Unknown	I/FKG 50	Unknown
15240	A-1	VD+UO	28/01/43	I/FKG 50	Enemy action
15241	A-1*	VD+UP	20/01/43	I/FKG 50	Landing accident
15242	A-1	VD+UQ	17/01/43	I/FKG 50	Enemy action
15243	A-1	VD+UR	Unknown	I/FKG 50	Unknown
15244	A-1	VD+US	26/06/43	ESt 177, I/FKG 50	Unknown
15245	A-1	VD+UT	??/05/43	I/FKG 50	Unknown
15246	A-1	VD+UU	??/05/43	I/FKG 50	Unknown
15247	A-1	VD+UV	Unknown	I/FKG 50, I/KG 40	Crashed (EF)
15248	A-1	F8+HU	Unknown	10./KG 40	Unknown
15249	A-1	VD+HX	Unknown	I/FKG 50	Unknown
15250	A-1	VD+HY	Unknown	Unknown	Crashed
15251	A-1	VF+RA	Unknown	I/FKG 50	Unknown
15252	A-1*	E8+FK	13/01/43	I/FKG 50	Crashed
15253	A-1*	VF+RC	??/05/43	Trainer	Unknown
15254	A-1*	VF+RD	11/02/44	V20, DVL	Unknown
15255	A-1	VF+RE	??/05/43	I/KG 4	Unknown
15256	A-1	VF+RF	21/02/44	EK 25	Enemy action
15257	A-1	VF+RG	Unknown	I/FKG 50	Unknown
15258	A-1	5J+CK	23/10/43	I/KG 100	Force-landed
15259	A-1	VF+RI	28/11/42	I/KG 40	Crashed (PE)
15260	A-1	VF+RJ	??/05/43	FFS (B) 16	Unknown
15261	A-1	VF+RK	??/05/43	I/KG 4	Unknown
15262	A-1	VF+RL	21/02/44	EK 25	Enemy action
15263	A-1	VF+RM	27/01/43	I/FKG 50	Crashed
15264	A-1	VF+RN	??/05/43	I/KG 40	Unknown
15265	A-1	VF+RO	12/08/43	I/KG 40	Unknown
15266	A-1	VF+RP	??/05/43	Trainer	Unknown
15267	A-1	VF+RQ	Unknown	Trainer	Unknown
15269	A-1	VF+RR	23/01/43	I/KG 40	Crashed (EF)
15270	A-1	VF+RS	??/05/43	Trainer	Unknown
15271	A-1	VF+RT	??/06/44	DLH	Engine fire
15272	A-1	VF+RU	02/05/44	I/FKG 50, FFS (B) 31	Unknown

Werk-Nummer	Model	Codes	Last Mentioned	User/Use	Fate
15273	A-1	VF+RV	??/05/43	FFS (B) 31	Unknown
15274	A-1	F8+OV	24/04/44	11./KG 40	Unknown
15275	A-1	VF+RX	07/05/44	11./KG 40	Engine failure
15276	A-1	F8+NV	02/04/44	11./KG 40	Unknown
15277	A-1	BF+TB	Unknown	KdE (A-3 pattern)	Unknown
15278	A-1	BF+TC	12/06/44	FFS (B) 31	Crash-landed
15279	A-1	BF+TD	??/05/43	I/KG 40	Unknown
15280	A-1	5J+EH	21/11/43	1./KG 4	Engine failure
332101	A-1*	VF+QA	13/11/43	V19 (DB 610s)	Crashed
332102	A-1*	VF+QB	02/03/44	V21 (*Kehl*)	Unknown
332103	A-1	VF+QC	05/05/43	1st *Kehl* a/c	Unknown
332104	A-1*	VF+QD	12/01/43	V22, KdE (to A-5)	Engine failure
332106	A-1	VF+QF	??/01/43	EHAG	Unknown
332109	A-1	VF+QI	02/06/43	Unknown	Total loss
332110	A-1	VF+QJ	30/01/43	Unknown	Engine failure
332111	A-1	VF+QK	18/04/44	FlUGr 1	Unknown
332112	A-1*	VF+QL	??/06/44	V23 (DB 610s)	Unknown
332121	A-1*	??+??	01/02/44	V36, KdE	Crash-landed
135006	A-3	ND+SA	15/06/44	FFS (B) 16	Crashed
135007	A-3	ND+SB	Unknown	*Luftwaffe*	Unknown
135008	A-3	ND+SC	28/01/43	ESt 177	Unknown
135009	A-3	ND+SD	Unknown	*Luftwaffe*	Unknown
135010	A-3	ND+SE	02/06/44	ESt 177	Crashed
135011	A-3	ND+SF	Unknown	*Luftwaffe*	Unknown
135012	A-3	ND+SG	Unknown	*Luftwaffe*	Unknown
135013	A-3	ND+SH	Unknown	*Luftwaffe*	Unknown
135014	A-3	ND+SI	Unknown	*Luftwaffe*	Unknown
135015	A-3	ND+SJ	Unknown	*Luftwaffe*	Unknown
135016	A-3	ND+SK	28/06/43	ESt 177	CTO
135018	A-3	ND+SM	Unknown	V30	Unknown
135020	A-3	ND+SO	Unknown	LT trials	Unknown
135024	A-3	ND+SS	06/11/43	V24, KdE	Unknown
332143	A-3	VD+XS	21/12/43	4./KG 100	Force-landed
332146	A-3	VD+XV	15/06/44	5./KG 1	Collision
332147	A-3	V4+IV	26/06/44	10./KG 1	Force-landed
332154	A-3	??+??	22/04/44	12./KG 100	Crash-landed
332157	A-3	??+??	Unknown	I/KG 1	Crash-landed
332169	A-3	??+??	22/01/44	I/KG 100	Unknown
332187	A-3	??+??	??/01/44	I/KG 40	U/C failure
332189	A-3	5J+AL	25/02/44	3./KG 100	Fighter kill
332193	A-3	??+??	??/01/44	I/KG 40	DWT
332195	A-3	??+??	05/03/44	II/KG 40	Enemy action
332198	A-3	5J+DL	22/01/44	3./KG 100	Crashed
332200	A-3	??+??	21/03/44	I/KG 100	Damaged
332201	A-3	??+??	30/01/44	I/KG 100	Engine fire
332203	A-3	??+??	14/11/43	III/KG 100	Crash-landed
332204	A-3	5J+AK	15/11/43	3./KG 100	Crashed
332206	A-3	5J+KK	02/03/43	2./KG 100	MIA
332209	A-3	??+??	30/01/44	I/KG 100	Engine fire
332210	A-3	5J+IL	20/02/44	3./KG 100	Force-landed
332212	A-3	5J+NK	08/01/44	2./KG 100	Crashed
332214	A-3	5J+RL	05/03/44	3./KG 100	Fighter kill
332216	A-3	??+??	02/01/44	I/KG 40	Crashed
332217	A-3	5J+HL	24/01/44	3./KG 100	AA kill

Werk-Nummer	Model	Codes	Last Mentioned	User/Use	Fate
332218	A-3	??+??	??/01/44	I/KG 40	DWT
332220	A-3	5J+IH	04/02/44	I/KG 100	Force-landed
332221	A-3	??+??	21/01/44	I/KG 100	Propeller damage
332222	A-3	5J+PK	24/02/44	2./KG 100	Operational loss
332224	A-3	6N+TL	22/03/44	I/KG 100	Fighter kill
332225	A-3	5J+VL	21/01/44	2./KG 40	Crash-landed
332226	A-3	??+??	22/12/43	I/KG 100	Unknown
332227	A-3	5J+QL	22/02/44	3./KG 100	MIA
332229	A-3	V4+PT	28/07/44	9./KG 1	Enemy action
332230	A-3	6N+WL	??/10/44	3./KG 100	Unknown
332231	A-3	5J+ZL	22/01/44	2./KG 40	Operational loss
332232	A-3	TM+IO	05/01/44	2./KG 40	Fighter kill
332235	A-3	6N+HK	22/04/44	2./KG 100	Crashed
332237	A-3	??+??	27/06/44	EK 25	Wrecked
332241	A-3	SL+WW	28/01/44	I/KG 100	Crashed
332251	A-3	5J+DH	18/02/44	I/KG 100	Crashed
332351	A-3	V4+CN	28/07/44	8.KG 1	Enemy action
332355	A-3	6N+GK	Unknown	2./KG 100	AA kill
332357	A-3	6N+IK	20/04/44	2./KG 100	Operational loss
332364	A-3	KP+PN	21/05/44	I/KG 1	Enemy action
332365	A-3	KP+PO	21/03/44	I/KG 100	Unknown
332366	A-3	V4+EN	27/06/44	5./KG 1	Engine failure
332367	A-3	KP+PQ	??/10/44	2./KG 100 (6N+EK)	Unknown
332375	A-3	6N+OK	19/03/44	2./KG 100	Operational loss
332379	A-3	6N+AK	19/04/44	2./KG 100	Operational loss
332385	A-3	??+??	24/03/44	I/KG 100	Unknown
332389	A-3	??+??	10/05/44	I/KG 1	U/C failure
332394	A-3	6N+BM	08/05/44	I/KG 100	Crashed
332401	A-3	??+??	16/04/44	I/KG 1	Landing accident
332407	A-3	??+??	16/04/44	10./KG 100	Ground contact
332408	A-3	??+??	10/05/44	I/KG 1	Pilot error
332410	A-3	??+??	13/04/44	I/KG 1	Instrument failure
332444	A-3	F8+EP	26/11/43	6./KG 40	Unknown
332471	A-3	V4+GH	20/03/44	I/KG 1	Crashed
332473	A-3	D7+BK	Unknown	Unknown	Unknown
332474	A-3	V4+LP	27/06/44	6./KG 1	Engine failure
332475	A-3	V4+FL	01/04/44	3./KG 1	Crashed
332476	A-3	??+??	Unknown	I/KG 1	Wrecked
332477	A-3	??+??	08/04/44	I/KG 1	Engine failure
332478	A-3	??+??	17/04/44	I/KG 1	Pilot error
332484	A-3	??+??	05/04/44	I/KG 1	Belly-landed
332491	A-3	V4+IS	01/08/44	8./KG 1	Wrecked
332495	A-3	V4+JS	01/08/44	8./KG 1	Crashed
332497	A-3	V4+HS	26/06/44	8./KG 1	Engine failure
332506	A-3	??+??	26/04/44	3./KG 100	Operational loss
332507	A-3	DW+CK	30/04/44	I/KG 100	Crashed
332511	A-3	V4+BK	29/04/44	2./KG 1	Ground contact
332515	A-3	DW+CS	03/05/44	I/KG 1	Wrecked
332518	A-3	DW+CV	15/06/44	4./KG 1	Crashed
332536	A-3	??+??	10/06/44	I/KG 1	Engine failure
332538	A-3	??+??	23/05/44	I/KG 1	Pilot error
332539	A-3	??+??	15/05/55	I/KG 1	Tyre defects
332540	A-3	V4+HL	18/07/44	3./KG 1	CTO
332543	A-3	V4+KL	28/06/44	3./KG 1	Ground contact

Werk-Nummer	Model	Codes	Last Mentioned	User/Use	Fate
332546	A-3	??+??	05/06/44	I/KG 1	Pilot error
332555	A-3	??+??	25/02/44	I/KG 100	Crashed
332610	A-3	V4+IN	28/07/44	5./KG 1	Engine failure
332618	A-3	6N+AS	Unknown	I/KG 100	Unknown
332628	A-3	??+??	Unknown	*Wekusta* 2/ObdL	Unknown
332629	A-3	??+??	Unknown	*Wekusta* 2/ObdL	Unknown
335001	A-3	??+??	24/06/44	V15	Ground contact
335002	A-3	??+??	Unknown	TS 2	DWT
335003	A-3	??+??	Unknown	DB trials	Crashed
335004	A-3	??+??	Unknown	V16 (trials)	Wrecked
335005	A-3	??+??	Unknown	V17 (trainer)	Wrecked
355055	A-3	??+??	03/03/43	KdE	Unknown
355056	A-3	??+??	22/07/43	V28, KdE	Unknown
355078	A-3	??+??	26/06/43	KdE	Unknown
535352	A-3	GP+WB	Unknown	Unknown	Unknown
535353	A-3	GP+WC	Unknown	V32	Unknown
535354	A-3	GP+WD	Unknown	V33	Unknown
535364	A-3	GP+WN	20/12/43	V34, KdE	Unknown
535365	A-3	GP+WO	23/08/43	KdE	Pilot error
535366	A-3	GP+WP	28/06/44	V35, KdE	Unknown
535367	A-3	F8+KM	26/11/43	4./KG 40	Operational loss
535369	A-3	F8+MM	26/11/43	4./KG 40	Operational loss
535370	A-3	GP+WU	29/11/43	I/KG 40	Unknown
535371	A-3	GP+WV	26/11/43	II/KG 40	Crash-landed
535372	A-3	F8+GH	12/09/43	4./KG 40	Unknown
535436	A-3	GP+WX	15/03/44	12./KG 100	Force-landed
535437	A-3	GP+WY	28/10/43	KdE	Unknown
535438	A-3	GP+WZ	??/01/44	II/KG 40	U/C failure
535439	A-3	GJ+RA	??/01/44	II/KG 40	Pilot error
535441	A-3	GJ+RC	??/02/44	II/KG 40	Unknown
535442	A-3	GJ+RD	10/11/43	II/KG 40	Force-landed
535443	A-3	F8+BN	21/11/43	5./KG 40	Operational loss
535444	A-3	F8+EP	26/11/43	II/KG 40	Operational loss
535445	A-3	GJ+RG	21/11/43	II/KG 40	Engine fire
535446	A-3	GJ+RH	15/09/43	KdE	Crashed (EF)
535447	A-3	F8+EN	23/01/44	5./KG 40	MIA
535448	A-3	GJ+RJ	23/01/44	II/KG 40	Enemy action
535454	A-3	GJ+RP	17/12/43	2./KG 40	Unknown
535457	A-3	GJ+RQ	27/03/44	II/KG 40	Air raid
535459	A-3	F8+LK	08/04/44	2./KG 40	Fighter kill
535460	A-3	GJ+RS	23/01/44	II/KG 40	Crashed
535533	A-3	NN+QA	??/01/44	II/KG 40	Crash-landed
535549	A-3	NN+QP	14/01/44	KdE	Crash-landed
535550	A-3	NN+QQ	30/08/44	V101 (B-5)	Scrapped
535551	A-3	NN+QR	21/11/43	II/KG 40	Landing accident
535552	A-3	NN+QS	??/09/43	KdE (to A-5)	Unknown
535553	A-3	NN+QT	26/11/43	II/KG 40	Landing accident
535554	A-3	NN+QU	17/12/43	II/KG 40	Crash-landed
535555	A-3	NN+QV	23/01/44	II/KG 40	Enemy action
535556	A-3	NN+QW	26/11/43	II/KG 40	Engine fire
535557	A-3	F8+IN	28/12/43	5./KG 40	MIA
535559	A-3	NN+QY	28/12/43	II/KG 40	Crash-landed
535560	A-3	NN+QZ	21/01/44	I/KG 100	Total loss
535561	A-3	??+??	??/01/44	II/KG 40	U/C failure

Werk-Nummer	Model	Codes	Last Mentioned	User/Use	Fate
535562	A-3	F8+LM	24/12/43	II./KG 40	Enemy action
535566	A-3	F8+IM	26/11/43	II./KG 40	Operational loss
535569	A-3	??+??	??/01/44	II./KG 40	MIA
535575	A-3	??+??	23/01/44	II./KG 40	Belly-landed
535670	A-3	F8+SK	09/06/44	2./KG 40	Crashed (EA)
535672	A-3	??+??	25/12/43	II./KG 40	Fighter victory
535673	A-3	F8+DX	08/06/44	12./KG 40	Unknown
535674	A-3	??+??	??/02/44	II./KG 40	Unknown
535675	A-3	??+??	Unknown	II./KG 40	Unknown
535677	A-3	F8+DM	26/11/43	II./KG 40	Operational loss
535678	A-3	F8+CV	27/03/44	11./KG 40	Unknown
535679	A-3	F8+PN	12/02/44	5./KG 40	Night-fighter kill
535680	A-3	F8+LH	29/02/44	I/KG 40	Crashed
535682	A-3	??+??	??/02/44	II./KG 40	Unknown
535683	A-3	??+??	21/11/43	II./KG 40	Crashed
535684	A-3	F8+BP	26/11/43	II./KG 40	Operational loss
535687	A-3	F8+KH	Unknown	1./KG 40	Unknown
535689	A-3	??+??	??/02/44	II./KG 40	Unknown
535690	A-3	F8+MK	26/03/44	1./KG 40	Unknown
535692	A-3	??+??	24/03/44	II./KG 40	DWT
535694	A-3	F8+GH	03/05/44	1./KG 40	Crashed (PE)
535695	A-3	??+??	??/02/44	II./KG 40	Fighter kill
535696	A-3	??+??	23/11/44	II./KG 40	Fighter kill
535730	A-3	CJ+FD	22/01/44	I/KG 40	Crashed (EF)
535731	A-3	F8+LK	08/06/44	2./KG 40	MIA
535732	A-3	CJ+FF	27/03/44	II./KG 40	Air raid
535733	A-3	F8+PV	28/01/44	11./KG 40	Unknown
535735	A-3	CJ+FI	23/01/44	II./KG 40	MIA
535736	A-3	CJ+FJ	21/01/44	I/KG 100	Unknown
535740	A-3	CJ+FN	20/02/44	II./KG 40	Belly-landed
535741	A-3	CJ+FO	21/01/44	1./KG 40	MIA
535743	A-3	CJ+FQ	21/01/44	1./KG 40	Night-fighter kill
535745	A-3	CJ+FS	21/01/44	1./KG 40	Crashed
535747	A-3	F8+HH	21/01/44	I/KG 40	Enemy action
535748	A-3	CJ+FV	29/01/44	1./KG 40	Crashed (PE)
535749	A-3	CJ+FW	24/03/44	I/KG 100	Belly-landed
535751	A-3	CJ+FY	??/02/44	I/KG 40	Unknown
535752	A-3	F8+AU	02/06/44	10./KG 40	Unknown
535753	A-3	??+??	23/01/44	II./KG 40	Unknown
535755	A-3	??+??	28/11/44	EK 36	Crashed
535758	A-3	F8+MK	22/03/44	2./KG 40	Unknown
535794	A-3	??+??	25/03/44	I/KG 40	Force-landed
535848	A-3	6N+AU	27/05/44	10./KG 100	Crashed
535849	A-3	GR+MG	Unknown	III/KG 100	Unknown
535850	A-3	GR+MH	28/06/44	V40, KdE	Unknown
535852	A-3	F8+QU	20/05/44	10./KG 40	Unknown
535854	A-3	GR+ML	27/03/44	II./KG 40	Air raid
535857	A-3	F8+AX	03/06/44	12./KG 40	Unknown
535862	A-3	GR+MT	10/03/44	10./KG 100	Flying accident
535865	A-3	GR+MW	Unknown	II./KG 40	Unknown
535866	A-3	GR+MX	Unknown	8./KG 100	Unknown
535868	A-3	GR+MY	27/03/44	II./KG 40	Air raid
535869	A-3	GR+MZ	??/01/44	II./KG 40	Unknown
535870	A-3	F8+TV	03/02/44	11./KG 40	Unknown
550001	A-5	6N+FM	09/10/44	4./KG 40	Unknown

Werk-Nummer	Model	Codes	Last Mentioned	User/Use	Fate
550002	A-5	KM+TB	08/05/45	V38, KdE	Remains captured at Prague-Rusiye
550003	A-5	KM+TC	22/02/44	4./KG 100	Collided
550004	A-5	6N+CM	09/10/44	4./KG 100	Unknown
550006	A-5	6N+DM	09/05/44	4./KG 100	Damaged
550031	A-5	6N+LM	15/05/44	4./KG 100	Crash-landed
550033	A-5	6N+EM	09/10/44	4./KG 100	Unknown
550034	A-5	6N+FM	17/05/44	4./KG 100	Enemy action
550035	A-5	KM+TK	30/06/44	KdE	Unknown
550036	A-5	KM+TL	26/06/44	V103, KdE	Scrapped
550038	A-5	KM+TN	25/05/44	V37, KdE	Unknown
550039	A-5	KM+TO	Unknown	II/KG 10	Unknown
550040	A-5	KM+TP	Unknown	II/KG 100	Unknown
550041	A-5	KM+TQ	Unknown	II/KG 40	DWT
550042	A-5	6N+IM	09/10/44	4./KG 100	Unknown
550043	A-5	6N+HM	09/10/44	4./KG 100	Unknown
550044	A-5	6N+HN	09/10/44	4./KG 100	Unknown
550045	A-5	6N+KM	09/10/44	4./KG 100	Unknown
550046	A-5	6N+BM	09/10/44	4./KG 100	Unknown
550047	A-5	6N+GM	??/10/44	EK 36	Unknown
550048	A-5	KM+TX	27/03/44	II/KG 40	Air raid
550049	A-5	KM+TY	02/05/44	V39, KdE	Unknown
550052	A-5	KM+UB	27/03/44	II/KG 40	Air raid
550054	A-5	KM+UD	09/10/44	6./KG 100	Unknown
550055	A-5	KM+UE	??/06/44	EHAG	Unknown
550056	A-5	KM+UF	??/05/44	II/KG 40	U/C failure
550057	A-5	KM+UG	??/09/44	II/KG 40	Unknown
550060	A-5	KM+UJ	Unknown	II/KG 40	Unknown
550061	A-5	KM+UK	Unknown	II/KG 40	Blown up
550065	A-5	KM+UO	21/03/44	5./KG 40	Air raid
550067	A-5	F8+BN	10/06/44	4./KG 40	Fighter kill
550068	A-5	KM+UR	27/03/44	II/KG 40	Night-fighter kill
550069	A-5	KM+US	27/03/44	II/KG 40	Air raid
550071	A-5	KM+UU	27/03/44	II/KG 40	Air raid
550072	A-5	6N+MM	09/10/44	4./KG 100	Unknown
550073	A-5	KM+UW	27/03/44	II/KG 40	Air raid
550074	A-5	KM+UX	10/06/44	5./KG 40	AA kill
550076	A-5	6N+LP	09/10/44	6./KG 100	Unknown
550079	A-5	F8+FM	02/05/44	4./KG 40	Crashed (EF)
550080	A-5	??+??	14/06/44	4./KG 40	MIA
550081	A-5	??+??	27/03/44	II/KG 40	Air raid
550082	A-5	??+??	??/06/44	I/KG 40	Crash-landed
550083	A-5	??+??	08/06/44	6./KG 40	Night-fighter kill
550084	A-5	??+??	27/03/44	Unknown	Air raid
550086	A-5	??+??	??/06/44	II/KG 40	U/C failure
550087	A-5	??+??	14/06/44	6./KG 40	Enemy action
550090	A-5	??+??	13/07/44	II/KG 40	Unknown
550098	A-5	??+??	14/06/44	4./KG 40	Night-fighter kill
550117	A-5	F8+HN	07/06/44	5./KG 40	MIA
550120	A-5	6N+AN	09/10/44	5./KG 100	Unknown
550121	A-5	6N+HP	09/10/44	6./KG 100	Unknown
550122	A-5	6N+BN	09/10/44	5./KG 100	Unknown
550123	A-5	F8+CU	26/05/44	10./KG 40	Unknown
550125	A-5	6N+IP	09/10/44	6./KG 100	Unknown
550127	A-5	6N+CC	09/10/44	II/KG 100	Unknown

Werk-Nummer	Model	Codes	Last Mentioned	User/Use	Fate
550128	A-5	6N+CN	09/10/44	5./KG 100	Unknown
550129	A-5	F8+DX	20/06/44	11./KG 40	Unknown
550130	A-5	6N+MM	09/10/44	4./KG 100	Unknown
550131	A-5	6N+DN	09/10/44	5./KG 100	Unknown
550132	A-5	6N+EN	09/10/44	5./KG 100	Unknown
550133	A-5	6N+AC	09/10/44	II/KG 100	Unknown
550134	A-5	6N+FN	09/10/44	5./KG 100	Unknown
550135	A-5	6N+MN	09/10/44	5./KG 100	Unknown
550136	A-5	6N+HN	09/10/44	5./KG 100	Unknown
550137	A-5	6N+IN	09/10/44	5./KG 100	Unknown
550138	A-5	6N+KN	09/10/44	5./KG 100	Unknown
550139	A-5	6N+GN	09/10/44	5./KG 100	Unknown
550140	A-5	6N+LM	09/10/44	5./KG 100	Unknown
550141	A-5	6N+BC	14/05/44	II/KG 100	Crashed (PE)
550142	A-5	??+??	??/05/44	II/KG 40	Crashed (icing)
550145	A-5	??+??	Unknown	II/KG 40	Unknown
550146	A-5	??+??	14/06/44	4./KG 40	Unknown
550147	A-5	F8+LP	??/08/44	II/KG 40	Blown up
550149	A-5	??+??	??/08/44	II/KG 40	Blown up
550150	A-5	6N+FP	09/10/44	6./KG 100	Unknown
550151	A-5	6N+KP	09/10/44	6./KG 100	Unknown
550153	A-5	??+??	24/05/44	II/KG 40	Ditched
550155	A-5	??+??	??/08/44	II/KG 40	Blown up
550157	A-5	??+??	??/05/44	II/KG 40	Air raid
550158	A-5	6N+DM	09/10/44	4./KG 100	Unknown
550159	A-5	6N+CP	09/10/44	6./KG 100	Unknown
550160	A-5	6N+EP	09/10/44	6./KG 100	Unknown
550161	A-5	6N+LN	09/10/44	5./KG 100	Unknown
550162	A-5	6N+BP	09/10/44	6./KG 100	Unknown
550166	A-5	6N+GP	09/10/44	6./KG 100	Unknown
550168	A-5	F8+AX	14/05/44	13./KG 40	Unknown
550170	A-5	6N+AP	09/10/44	6./KG 100	Unknown
550172	A-5	6N+HP	09/10/44	6./KG 100	Unknown
550173	A-5	??+??	09/07/44	4./KG 40	Fighter kill
550175	A-5	F8+JH	10/06/44	1./KG 40	MIA
550191	A-5	F8+BB	23/05/44	I/KG 40	Engine failure
550195	A-5	??+??	05/07/44	6./KG 40	Night-fighter kill
550197	A-5	F8+KK	08/06/44	2./KG 40	MIA
550198	A-5	F8+BK	07/06/44	2./KG 40	Fighter kill
550199	A-5	F8+DH	10/06/44	1./KG 40	MIA
550202	A-5	F8+KH	??/08/44	V31, II/KG 40	Blown up
550203	A-5	TM+IG	05/07/44	6./KG 40	Night-fighter kill
550204	A-5	F8+FK	07/06/44	2./KG 40	Fighter kill
550206	A-5	F8+MH	07/06/44	1./KG 40	MIA
550210	A-5	??+??	05/07/44	4./KG 40	MIA
550211	A-5	F8+MK	08/06/44	2./KG 40	MIA
550213	A-5	??+??	05/07/44	5./KG 40	Unknown
550215	A-5	F8+FH	13/06/44	1./KG 40	MIA
550220	A-5	??+??	??/08/44	II/KG 40	Blown up
550221	A-5	??+??	13/06/44	6./KG 40	Wrecked
550225	A-5	??+??	??/08/44	II/KG 40	Blown up
550230	A-5	GP+RA	13/07/44	I/KG 40	Unknown
550235	A-5	GP+RE	Unknown	I/KG 40	Unknown
550255	A-5	GP+RX	05/07/44	6./KG 40	Enemy action

Werk-Nummer	Model	Codes	Last Mentioned	User/Use	Fate
550256	A-5	GP+RY	mid-1945	Rebuilt to A-7	War booty
550257	A-5	GM+DA	07/07/44	6./KG 40	CTO
550258	A-5	6N+MP	27/07/44	6./KG 100	Crash-landed
550263	A-5	6N+AM	09/10/44	4./KG 100	Unknown
550266	A-5	GM+DJ	09/10/44	6./KG 100	Unknown
550319	A-5	DV+OL	09/10/44	6./KG 100	Unknown
550324	A-5	DV+OQ	09/10/44	6./KG 100	Unknown

*He 177A-1s known to have been converted to A-3s.

NB: This list does not cover all He 177s built

AA	Anti-Aircraft (fire)
CTO	Crashed on Take-Off
DLH	Deutsche Luft Hansa
DWT	Damaged While Taxying
EA	Enemy Action
EF	Engine Failure
EHAG	Ernst Heinkel AG
EK	*Erprobungskommando* (Proving/Test Command)
FlUGr	*Flugzeugüberführungsgruppe* (Aircraft Ferry Group)
PE	Pilot Error
MIA	Missing In Action
TOA	Take-Off Accident
TS	*Technische Schule* (Technical School)
U/C	Undercarriage

Bibliography (a selection)

Baumbach, W.	*Zu Spät?*, Stuttgart, 1977
Bekker, C.	*Angriffshöhe 4000*, Hamburg, 1964
Bowyer, M.J.F.	*Air Raid*, Wellingborough, 1986
Brembach, H.	*Adler über See*, Oldenburg, 1962
Cross, R.	*The Bombers*, New York, 1987
Dierich, W.	*Die Verbände der Luftwaffe 1935-1945*, Stuttgart, 1976
Douhet, G.	*Il dominio dell'aria*, 1936
Feuchter, G.W.	*Geschichte des Luftkriegs*, Bonn, 1954
Gersdorff, K. von	*Flugmotoren und Strahltriebwerke*, 1981
Götsch, E.	*Henschel-Flugzeuge, Sonderdruck d. Henschel-Werke*, Kassel, 1988
Green, W.	*Warplanes of the Third Reich*, London, 1972
Griehl, M.	*Do 217/317/417*, Stuttgart, 1987
Griehl, M.	*Junkers Bombers, Vol 1*, London, 1987
Hahn, F.	*Deutsche Geheimwaffen 1939-1945*, Heidenheim, 1963
Hitchcock, Th.H.	*Junkers Ju 288*, 1974
Hitchcock, Th.H.	*Junkers Ju 290*, 1975
Hümmelchen, G.	*Die Deutschen Seeflieger 1939-1945*, Munich, 1976
Kurowski, F.	*Seekrieg aus der Luft*, Herford, 1979
Kurowski, F.	*Luftwaffe über Russland*, Rastadt, 1987
Lange, B.	*Das Buch der Deutschen Luftfahrttechnik, Bd 1 & 2*, 1970
Lauck, F.	*Der Lufttorpedo*, Munich, 1981
Nowarra, H.J.	*Die Deutsche Luftrüstung 1933-1945, Bd 1 bis 4*, 1985 ff
Nowarra, H.J.	*Ju 88*, Stuttgart, 1978
Philpot, B.	*German Bombers over Russia*, Cambridge, 1979
Poturyn, F. von	*Luftmacht*, Heidelberg/Berlin, 1938
Price, A.	*Blitz über England*, Stuttgart, 1978
Price, A.	*Pictorial History of the Luftwaffe 1933-1945*, London, 1969
Price, A.	*German Air Force Bombers of World War II, Vol II*, 1969
Price, A.	*Heinkel He 177 'Greif'*, Windsor, 1975
Rowehr, J. & Hümmelchen, G.	*Chronik des Seekriegs 1939-1945*, Oldenburg & Hamburg, 1968
Rust, K.C.	*Fifteenth Air Force Story*, 1976
Schliephake, H.	*Flugzeugbewaffnung*, Stuttgart, 1977
Schliephake, H.	*Wie die Luftwaffe wirklich entstand*, Stuttgart, 1972
Trenkle, F.	*Bordfunkgeräte*, Koblenz, 1986
Trenkle, F.	*Die Deutsche Funk-Navigations- und Funkführungsverfahren bis 1945*, Stuttgart, 1945
Wolf, W.	*Luftangriff auf die Deutsche Industrie*, Munich, 1985
West, K.S.	*The Captive Luftwaffe*, London, 1978

Photo Credits Balke (11), BMW (2), Bekker (7), Birkholz (1), Bundesarchiv (15), Chapman (1), Charles (1), Creek (16), Deutsches Museum München (2), Dornier (2), DVLR (1), Fleuret (1), Forschungsgruppe Luftfahrtgeschichte e. V. (6), Heck (2), Heinkel (21), Herrendorf (1), Herwig (8), Kruse (5), Lange (1), Lutz jr. (13), van Mol (7), Müller-Romminger (3), Marchand (6), Menke (3), Nowarra (10), Ott (3), Pause (1), Petrick (8), Radinger (4), Roosenboom (36), Schliephake (8), Schirer (1), Selinger (6), Sengfelder (2), Thomas (1), Zucker (14), Verfasser (10).

Glossary

AEG:	Allgemeine Elektrizitat Gesellschaft
AG:	*Aktiengesellschaft* (Joint-Stock Company)
AGr:	*Aufklärungsgruppe* (Reconnaissance Wing)
ASV:	Anti-Surface Vessel (radar)
AufklGr(F):	*Fernaufklärungsgruppe* (Long-Range Reconnaissance Wing)
AWB:	Arado-Werke at Brandenburg
(B) *Schule*:	Advanced Flight Training School/Blind Flying School
BAL:	*Bauaufsicht Luft* (Construction Supervisor, Air)
BehBG/KG:	*Behelfsbombergeschwader* (Auxiliary Bomber Group)
BL:	*Bugstandlafette* (Nose Position Gun-Mount)
Blitzkrieg:	Lightning War
BMW:	Bayerische Motoren Werke (Bavarian Motor Works)
BV:	Blohm und Voss
C-Amt:	Technical Office of the *Luftfahrtkommissariat*
(C) *Schule*:	Flight Training School (Multi-Engined Aircraft)
DB:	Daimler-Benz
Deutsche Reichsbahn:	German State Railway
DFS:	*Deutsche Forschungsanstalt für Segelflug* (German Research Institute for Sailplanes)
Dipl-Ing:	*Diplomingenieur* (Academically-Qualified Engineer)
DL:	*Drehlafette* (Swivel Gun-Mount)
DLH:	Deutsche Luft Hansa (German State Airline; shortened to Lufthansa on 1 January 1934)
Do:	Dornier
DVL:	*Deutsche Versuchsanstalt für Luftfahrt* (German Aviation Experimental Establishment)
E:	*Erprobungs* (Test/Proving)
EHAG:	Ernst Heinkel AG
EHF:	Ernst Heinkel *Flugzeugwerke*
EK:	*Erprobungskommando* (Proving/Test Command)
E-Lehr Kdo:	*Erprobungs und Lehr Kommando* (Test and Instruction Detachment)
Erg-KGr:	*Ergänzungs-Kampfgruppe* (Reserve/Replacement Training Bomber Wing)
E-Staffel:	*Erprobungsstaffel* (Proving/Test Squadron)
E-Stelle:	*Erprobungsstelle* (Proving/Test Centre)
ETC:	*Elektrische Trägervorrichtung für Cylinderbomben* (Electrically-Operated Carriers for Cylindrical Bombs)
FAGr:	*Fernaufklärungsgruppe* (Long-Range Reconnaissance Wing)
FDL:	*Fernbedienbare Drehlafette* (Remote-Controlled Rotating Gun-Mount)
Feldwebel:	Sergeant-Major
Feldwerft-Abt zbV:	*Feldwerftabteilung zur besonderen Verwendung* (Field Repair Detachment for Special Duties)
FFS (B):	*Flugzeugführerschule* B (Advanced Flight Training School/Blind Flying School)

FFS (C):	*Flugzeugführerschule* C (Flight Training School, Multi-Engined Aircraft)
FHL:	*Fernbedienbare Hecklafette* (Remote-Controlled Rear Gun-Mount)
FKG:	*Fernkampfgeschwader* (Long-Range Bomber Group)
FL:	*Ferngesteuert Lafette* (Remote-Controlled Gun-Mount)
Flak:	*Fliegerabwehrkanone* (Anti-Aircraft Cannon)
Fliegerdivision:	Air Division
Fliegerführer:	Air Commander
Flieger-Ing:	Airman-Engineer
Fliegerkorps:	Air Corps
FuG:	*Funkgerät* (Radio/Radar Set)
FuNG:	*Funknavigationsgerät* (Radio/Radar Navigation Set)
Fw:	Focke-Wulf
Gefreiter:	Leading Aircraftsman
General	Air Marshal
GdF:	*General der Flieger* (General of Fighters)
GdJ:	*General der Jagdflieger* (Commanding General, Fighter Branch)
GdK:	*General der Kampflieger* (General of Bombers)
Generalfeldmarschall:	Marshal
General-Ing:	General Engineer
Generalleutnant:	Air Vice-Marshal
Generalmajor:	Air Commodore
Generaloberst:	Air Chief Marshal
Generalstabs-Ing:	Engineer of the General Staff
GenTT:	*General der Truppentechnik* (General of Aircraft Maintenance)
Geschwader:	Group
GL/C:	*Generalluftzeugmeister, Technisches Amt* (General of Air Production, Technical Office)
GLZ:	*Generalluftzeugmeister* (Chief of Air Force Supply and Procurement)
GM:	Glykol-Methanol
GmbH:	*Gesellschaft mit beschränkter Haftung* (Limited-Liability Company)
Gruppe(n):	Wing(s)
Gruppenkommandeur:	Officer commanding a Wing
Hauptmann:	Flight Lieutenant
HDL:	*Hydraulische Drehlafette* (Hydraulically-Operated Rotating Gun-Mount)
HL:	*Hecklafette* (Tail Gun-Mount)
Hs:	Henschel Flugzeugwerke AG
HWO:	Heinkel-Werke Oranienburg
HZ:	*Hohenlader Zentrale* (Built-In High-Altitude Compressor)
I:	Illing
IFR:	In-Flight Refuelling
Ing:	*Inginieur* (Engineer)
Jägerstab:	Fighter Staff
JFM:	Junkers Flugzeug- und Motorenwerke
JGr:	*Jagdgruppe* (Fighter Wing)
Kampfgeschwader:	Bomber Group
Kampfgruppe(n):	Bomber Wing(s)
KdE:	*Kommandeur der Erprobungsstellen* (Commander of Proving/Test Establishments)
KG:	*Kampfgeschwader* (Bomber Group)
KGrzbV:	*Kampfgruppe zur besonderen Verwendung* (Battle Group for Special Duties/Transport Wing)
Kriegsmarine:	German Navy
Leutnant:	Pilot Officer
LFA:	

LgKdo:	*Luftgaukommando* (Air Zone Command)
LMA:	*Luftmine* (Aerial Mine) Type A
LMB:	*Luftmine* (Aerial Mine) Type B
Lotfe:	*Lofternrohr* (Telescopic Bomb-Sight)
LT:	*Lufttorpedo* (Aerial Torpedo)
LTS:	*Lufttransportstaffel* (Air Transport Squadron)
Luftflotte:	Air Fleet
Luftflottenkommando:	Air Fleet Command
Luftwaffenführungsstab:	Air Force Operations Staff
Luftwaffengeneralstab:	Air Force General Staff
Major:	Squadron Leader
Me:	Messerschmitt
MG:	*Maschinengewehr* (Machine-Gun)
MK:	*Maschinenkanone* (Machine-Cannon)
ObdL:	*Oberbefehlshaber der Luftwaffe* (Commander-in-Chief Air Force)
Oberleutnant:	Flying Officer
Oberst:	Group Captain
Oberstleutnant:	Wing Commander
OKL:	*Oberkommando der Luftwaffe* (Air Force High Command)
OKW:	*Oberkommando der Wehrmacht* (Armed Forces High Command)
PaK:	*Panzerabwehrkanone* (Anti-Tank Cannon)
PC:	*Panzerbrechende Bombe, Cylinderisch* (Cylindrical Armour-Piercing Bomb)
PeilG:	*Peilgerät* (Direction-Finding Set)
Reichsmarschall:	State Marshal
Revi:	*Reflexvisier* (Reflex Bomb-Sight)
RLM:	*Reichsluftfahrtministerium* (State Ministry of Aviation)
RM:	*Reichmark*
RMfRuK:	*Reichsministerium für Rustungskraft* (State Ministry of War Supply)
SC:	*Sprengbombe, Cylindrisch* (Cylindrical Fragmentation Bomb)
SD:	*Sprengbombe, Dickwandig* (Thick-Walled Fragmentation Bomb)
SG:	*Sondergerät* (Special Device/Equipment)
st:	static thrust
Stab:	Staff
Staffel(n):	Squadron(s)
Stuka:	*Sturzkampffleugzeug* (Dive-Bomber; specifically the Ju 87)
Stuvi:	*Sturzkampfvisier* (Dive-Bombing Sight)
Süd:	South
Technische Amt:	Technical Office of the RLM
TK:	*Turbo Kompressor*
USAAF:	United States Army Air Force
V:	*Versuchs* (Experimental)
Verkehrsinspektion der DLH:	Traffic Inspectorate of the German State Airline
VfH:	*Versuchsstelle für Höhenflüge* (Experimental Establishment for High-Altitude Flights)
Wekusta:	*Wetterkungdungstaffel* (Weather Reconnaissance Squadron)
WFG:	Weser Flugzeugbau GmbH
Wk-Nr(n):	*Werk Nummer(en)* (Work Number(s))
WL:	*Walzenlafette* (Roller Gun-Mount)
Z:	*Zwilling* (Twin)
Zerstörer:	Destroyer/Heavy Fighter aircraft

Index

I/FKG 50 operations 74–5, 77–83, 92
I/KG 40 operations 126–9
I/KG 100 operations 137–44
II/KG 40 operations 129–36
II/KG 100 operations 145–51

Aalborg airfield 145, 147, 149–50
aerial torpedoes 104, 105–8
'Amerika Bomber' 183–191
anti-shipping tasks 59, 105–8, 114, 118–19, 123–4, 130–1, 136
Arado works 62, 64, 65, 70, 71, 165
armament, defensive 30–42, 95–6, 98, 102, 166, 168, 170–1, 180–2, 203
(*see also* cannon installations; dorsal, nose and ventral gun positions)

BMW 801 engines 98, 101
bomb-bays 43–5
bomb-loads 42–9, 62, 177, 182, 184
bomb-sights 43, 44, 180
bombing raids, first 49
Bordeaux-Mérignac airfield 126–7, 130–1, 133–6
Brandis airfield 84, 85

cannon installations 35, 83, 106, 109, 110, 111, 166, 187
Châteaudun airfield 125, 126–7, 142
cockpit configurations 162
cockpits, pressurised 176–80
comparison flight tests 58–9

D-Day Invasion 134, 136, 203
Daimler-Benz 104, 217
 DB 603 engines 98, 101, 164, 167, 180
 DB 606 engines 14, 22, 26, 54, 60, 92–3, 217, 224
 DB 610 engines 52–4, 62, 66, 94, 101, 164, 180, 224
 DB 613 engines 69, 97, 98, 224
defensive armament 30–42, 95–6, 98, 102, 166, 168, 170–1, 180–2, 203
(*see also* cannon installations; dorsal, nose and ventral gun positions)
Dornier
 Do 11: 2
 Do 17: 4
 Do 19: 5, 7
 Do 217: 114, 115, 118, 124
 Do 217P 200, 201
 Do 335 *Pfiel* 71, 172
dorsal gun positions 33–5, 38, 39, 95–6, 166, 168, 170–1, 180–2
drop-loads 42–9, 62, 177, 182, 184
drop tanks 170, 171
Düppel ('Window') 127, 140

E-Staffel (Operational Trials Unit) 177: 29, 57, 58, 59, 60, 62
E-Stelle (Proving/Test Centre)
 Karlshagen 115, 118, 123
 Peenemünde 115, 118, 119–20
 Rechlin 1, 23, 46, 52, 56, 62, 109
 Tarnewitz 23, 33, 36, 39, 41
England, raids on 129, 139–44

Farman works 192–4, 203–5
Farnborough, RAE 211, 215
Fassberg airfield 144, 147, 148, 150
flight tests, comparison 58–9
Focke-Wulf
 Fw 190A-8: 112
 Fw 200 Condor 45–6, 123–4
 Fw 300: 187
 Ta 400: 187–8, 196, 197–8
four-engined configuration, proposed 97, 99–101, 157, 159–60, 177, 203
Francke, Dipl-Ing *Leutnant* Carl 15, 16, 17, 19, 162, 175, 197
French post-war service 205–7
Fritz X (PC1400X) glide bomb 114, 118, 119, 121–3, 143–4
fuselage construction 101, 103

Göring, *Reichsmarschall* Hermann 1, 3, 7, 24, 53, 66, 88–9, 104
Grosszerstörer (Big Destroyer), Heinkel He 177 A-3: 111
guided weapons 104, 105–8, 114–23
Günter, Prof Dipl-Ing Siegfried 96–7, 219

Heidfeld-Schwechat plant, raids on 170
Heinkel, Prof Dr-Ing Ernst 29, 53, 56, 96, 170, 177, 193, 197
Heinkel
 He 111: 3, 4, 217
 He 119: 14
 He 162 *Salamander* 172–3
 He 177
  authorisation to build 9
  prototypes commenced 16
  V1: 17, 25, 26, 220, 225, 232
  V2: 18–20, 220, 225, 232
  V3: 19–20, 220, 225, 232
  V4: 22, 220, 225, 232
  V5: 21, 26, 220, 225, 232
  V6: 21–2, 23, 26, 33, 45, 220, 225, 232
  V7: 22–4, 27, 30, 56, 221, 225, 232
  V8: 23–4, 26, 221, 225, 232
  A-01: 23, 222, 227, 232
  A-02: 26, 33, 46, 50, 222, 232
  A-03: 26, 33, 34, 55, 222, 232
  A-04: 51, 222, 232
  A-05: 29, 51, 222, 232
  A-07: 29, 222, 232
  A-1: 35–6, 43, 48, 50, 52–4, 64–5, 222, 225–7, 232–5 (*see also Zerstörer*)
  A-3: 39, 41, 53–4, 60–1, 65–7, 94–5, 105–6, 111, 222, 225–7, 235–8
  A-5 'Spring Bomber' 70–1, 93, 96, 104, 133, 164, 171–2, 223, 226, 229, 238–41
  A-6: 96, 101, 223, 230
  A-7: 96, 157–8, 223, 230

B-5: 159–62, 165–7, 172, 230
  B-5 V101: 165, 166, 167, 169
  B-5 V102: 163, 166, 167, 169
  B-5 V103: 166, 167, 170
  B-7: 160–3, 167–8, 172, 231
  post-war fate 207, 210–16
He 177H 177–3
He 274: 169, 177–83, 191, 203–9
  AAS 01A 203–9
  AAS 01B 203, 205–7
  post-war use 205–9
He 277: 195–9, 202–3
He 343 jet bomber 187
Henschel
  Hs 130: 199–200, 201
  Hs 293 guided missile 114–15, 117, 119–23, 130, 131
  Hs 294 bomb-torpedo 117–18, 121–3
Hertel, Dipl-Ing Heinrich 9, 14, 54, 66–7, 92, 96, 165, 217
high-altitude bomber 176–83
'Hindenburg', KG/I operations 83–91
Hitler, Adolf 69, 92
hydraulic devices 126

Junkers (JFM)
  He 177 version 96
  Ju 52/3m 2
  Ju 86: 3
  Ju 88: 11, 13
  Ju 89: 6, 7
  Ju 90: 7
  Ju 186: 200
  Ju 287: 203
  Ju 288: 165
  Ju 290: 188–91
  Ju 390: 190–2
  Ju 488: 199, 201–2

KG/I 'Hindenburg' operations 83–91
KG 4: 137–8
KG 40, service with 123–37, 213–14
KG 100 'Wiking', service with 126, 128–9, 134, 137–51, 152–3
Kehl radio-guidance equipment 42, 94, 96, 115–17, 119–20, 122–3, 125, 133, 145
Knemeyer, *Oberstleutnant*, 167, 199, 201
Kramer, Dr Max 118

London, raids on 129, 139–43
long-range reconnaissance 55, 151–6, 186

low-level bombing 89, 91
Lucht, General-Ing 8, 21

maritime warfare 59, 105–8, 114, 118–19, 123–4, 130–1, 136, 186
Mayerhofer, *Hauptmann* Bodo 145, 149
Messerschmitt
  Me 261: 185
  Me 262: 76, 172
  Me 264: 185–7, 202
mid-air refuelling 57–8
Milch, *Generalfeldmarschall* Erhard 1, 7, 12, 29, 66, 82–3, 114–15, 162, 164–5
Misteln (Mistletoe) 'piggy-back' composite weapons 173–5
Mons, Major 29, 58, 59, 60, 97–8, 106, 130

Neptun radar 111
nose gun positions 33–5, 42, 95–6, 98, 102, 166, 168, 170–1, 180–1

Operation *Eisenhammer* (Iron Cross) 175
Operation *Shingle* 130
Operation *Steinbock* (Capricorn) 127, 129, 139–43
operational trials 55–62

PC1400X (Fritz X) glide bomb 114, 118, 119, 121–3, 143–4
Peltz, General 169
Peterson, *Oberst* 29, 52, 59, 167
Poland, invasion of 13
Portsmouth, raid on 143–4
pressurised cockpits 176–80
production 63–71
Project P1041: 8–9

RLM (State Ministry of Aviation) 1, 7–8, 11–13, 15, 31, 34, 42, 93–4, 182, 197
radar 50, 62, 111, 137, 155, 171, 210
radio equipment 50, 96, 182
radio-guidance equipment (*see Kehl*)
reconnaissance, long-range 55, 151–6, 186
refuelling, mid-air 57–8
remote-controlled rear gun-mounts 36, 39
remote-controlled rotating gun-mounts 33, 34, 35, 38, 39
Reper, *Oberleutnant*, 126–7
Rieder, *Hauptmann*, 131, 133
rocket launchers 111

Roma, Italian battleship, attack on 118–19

Schede, Major 77, 79, 109
Schleppgerät (Towed Device) 112–14
Schlosser, *Hauptmann*, 79, 83, 92–3
Schubert, Major 218
Second World War begins 13
servicing 61
Spanish Civil War 3–4
'Spring Bomber', He 177 A-5: 70–1, 93, 96, 104, 133, 164, 171–2, 223, 226, 229, 238–41
Stalingrad, raids on 77, 79–83
Sud-Ouest SO 4000: 207, 209

tail gun positions 31–41, 96, 98, 167–8, 203
tanks, attacks on 89, 91
target-released weapon-triggering devices 111–12, 114
tests, comparison flight 58–9
torpedoes, aerial 104, 105–8
Towed Device (*Schleppgerät*) 112–14
training 73–6
trials, operational 55–62
twin fin/rudder assembly 97, 163–4, 166–7

Udet, *Oberst* Ernst 7, 9, 16, 29, 217
undercarriage 68, 97, 135

Velikiye Luki area, raids on 87, 88
ventral gun positions 33–5, 39, 98, 102, 146, 168, 170, 180–2
von Kalckreuth, *Hauptmann* 138
von Riesen, *Oberstleutnant* Horst 87, 88
Vorwald, General 120, 201

weather reconnaissance 151–6
Wekustal/Obdl, service with 151–6
'Wiking', KG 100, service with 126, 128–9, 134, 137–51, 152–3
'Window' (*Düppel*) 127, 140
wing centre-section, proposed 97, 99–101
wing deformation 53
Zaporozhye airfield 77
Zerstörer (Destroyer) 58, 65, 83, 106, 109–11 (*see also* He 177 A-1/U-2)
Zwölfaxing plant, raids on 170